# Cognitive Optometry 2

*Exploring Human Awareness*

THRU   Vision

*2010*

# Nazir  Brelvi, OD

authorHOUSE®

*AuthorHouse™*
*1663 Liberty Drive*
*Bloomington, IN 47403*
*www.authorhouse.com*
*Phone: 1-800-839-8640*

*©2010 Nazir Brelvi OD. All rights reserved.*

*No part of this book may be reproduced, stored in a retrieval system, or transmitted by any means without the written permission of the author.*

*First published by AuthorHouse 1/19/2010*

*ISBN: 978-1-4490-6555-3 (e)*
*ISBN: 978-1-4490-6554-6 (sc)*

*Library of Congress Control Number: 2009914365*

*Printed in the United States of America*
*Bloomington, Indiana*

*This book is printed on acid-free paper.*

*For Sarah, Jeff & Jane*

*"Thinking about Truth is not enough. We must realize Truth ourselves, we must see things as they really are, not just as they appear to be. It is right understanding that is real Wisdom."*

**The Upanishads**
**1500 - 1200 BC**

# Contents

## Section 1 — 1
Introduction — 1
Awareness — 3
Evolution of the Nervous System — 5
Development of the Nervous System — 9
Primary Sensory Receptors: An Overview — 13
Sensory Neurons — 15
"Brains" — 16
Basic Architecture of "Brains" — 18
White Matter — 21
Myelination of the CNS & Cortical Plasticity — 22
The Corpus Callosum — 26
The evolving Cerebral Cortex (CC) — 28
The Prefrontal Cortex (PFC) — 33
Subjective Sensory Experience — 34
Multi-modal Integration — 35
Object Recognition — 38
Maturation of Subcortical Visual Circuits — 39
Development of "Face" Recognition — 40
Language? Acquisition — 41
Brain Plasticity & Critical Periods — 42
Cognitive Psychology — 45
The Brain's 'Tracking' System — 47

## Section 2 — 50
Sensory Receptive Fields — 50
Brain Maps & Visual Space — 52
Independent or Convergent Pathways — 53
Neuronal Gestalts? — 54
Neuro-matrix — 56
Neuro-Magic: "Doors" of Perception — 59
Ensemble Coding or Re-awareness — 61

## Section 3 — 63
What is Vision? — 63
Temporal events in Vision — 67
Retinotopic Maps & Visual Fields — 71
Binocular Vision — 72
Fixational Eye Movements — 76
Fusion, Stereopsis & Panum's Space — 80
Binocular Rivalry — 83
Optical Illusions — 87
The Primary Visual Cortex or V1 — 90
Beyond V1 — 92
Visual Perception — 94

## Section 4 — 97
Neural Substrates of Awareness — 97
Memory & Awareness — 100
Emotions & Awareness — 103
The Brainstem — 106
Reticular Formation — 107
The Reticular Activating System (RAS) — 108
Some Specific BrainStem Nuclei — 110
Neuro-modulators — 112
Neuro-Peptides — 114
The Limbic System: A modern approach — 115

## Section 5 — 119
Animal 'Minds' — 119
Adaptation and the Brain — 121
Basic Instincts and "Intelligence" — 123
"Ghost" in the machine? — 126
Human "Intelligence" — 128
Somatic Marker Hypothesis — 131
Hemispheric Lateralization & Human Uniqueness — 134
Innate Speech — 140
Birth of Sentience — 143
Human Consciousness — 147

| | |
|---|---|
| Primary "Consciousness" or Heightened Awareness | 151 |
| Intuition & 'Gut' feelngs | 153 |
| Unity of Consciousness | 156 |
| The search for 'Qualia' | 160 |
| Closing Thoughts | 167 |

# Further Reading:     171

# About the Author     172

*"Although the content of consciousness depends in large measure on neuronal activity, awareness itself does not...*

*To me it seems more and more reasonable to suggest that the mind may be a distinct and different essence."*

Wilder Penfield

*1975*

# Preface

What you are about to read is a diverse compilation of thoughts, informed guesses and brief synopses of selected scientific discourses, that I have culled from published accounts found in the public arena in various "elite" magazines, scientific journals, Wikipedia and lay media.

The main purpose of this book is not to dissect, nit-pick, micro- analyze or punch holes in the diverse views expressed by numerous fellow scientists mentioned in it, but to present their thoughts and informed speculations in a unified context. For, after all, most of these mental 'tinkerers' have devoted literally decades of their lives to their chosen field of endeavor.

I have also tried to emphasize the ways in which much progress has been achieved in the various neuro-sciences and the creativity of the progress by which the best, most productive investigators have assembled disparate and seemingly contradictory pieces of information into unexpected scientific and therapeutic syntheses.

Awareness, the central topic of this book, especially human awareness is not only a vast topic to explore, but is still in the process of being defined as an "explorable" entity. There are numerous "theories" floating out there within informed circles, ranging from the simple to the bizarre. Various "camps" have also arisen from the neurological quagmire and once-solid boundaries between the real and virtual, the physical and metaphysical seem to merge imperceptibly into the nebulous.

My next book in the series, *Cognitive Optometry 3*, will discuss and explore the neuro-chemical landscape of the human brain: how neuro-ceuticals will affect our future generations via customized formulations of 'mind-altering' drugs, both of the legal and street variety. I will explore answers to questions emerging about why college students and executives habitually ingest stimulant or street-legal drugs to enhance routine mental performance and if there are any contra-indications to such "off-label" usage.

Just like my first Introductory book on *Cognitive Optometry*, almost everything within the twin covers of this book was compiled, edited, and written by me over the last *several* years, as I attended to a steady stream of patients at the office or sat quietly beside my *koi* pond at home, frequently accompanied by Suki, my little Shih-tzu companion.

What I have observed during this active search for "truth" in the neurobiological sciences is that most valuable insights into human awareness to emerge during this first decade of the new millennium, did not come from the disciplines traditionally concerned with "Mind" issues i.e. philosophy, psychology etc. but from a sort-of *merger*, if you will, of these disciplines with the so-called "Biology of the Brain".

However, I also noticed, that this 'merging' is not restricted to just the biological sciences but also evident in numerous and multiple areas of scientific research. Science, as practiced today, is no longer the exclusive domain of so-called scientists. It has become an integral part of modern life and contemporary culture, especially here in the United States.

Toward the middle of the 20th Century, it dawned on some scientists that the operations (computations) performed by computers were not unlike what a human being does when solving a problem. This notion was embraced by some far-sighted psychologists like Jerry Bruner and George Miller and the cognitive approach, which emphasized internal mechanisms that process information, was born.

Scientists acquire empirical evidence according to the kinds of questions they pose and the methods of inquiry they adopt. Until now, the questions and methods have been overwhelmingly objective and quantitative, which inevitably has produced an objective, quantitative view of the universe at large, including the 'mind', awareness and human cognition.

The challenge facing the brain sciences is to either discover the laws or basic principles of mental phenomena by careful examination, with as few ideological assumptions as possible (a contemplative approach) or continue exploring the mind primarily by examining its physical correlates, which only reinforces the materialistic assumptions held during the 19th century, when scientific study of the mind began.

The cognitive movement has had a tremendous impact on various scientific disciplines. Information-processing concepts were also adopted by workers in linguistics, anthropology, mathematics, physics and other social sciences. The fact that cognitive processes are not dependent on consciousness (which actually depends on so-called unconscious cognitive processes) means that the mental vs. physical dilemma does not have to be overcome in order to study the brain mechanisms of cognition.

As you read in my earlier book on *An Introduction to Cognitive Optometry*, many of the studies by cognitive scientists are also topics of research pursued by neuro-scientists. Led by breakthroughs in understanding the psychology of cognition, cognitive neuro-scientists have been very successful in relating perception, attention, memory, emotions and thinking to underlying mechanisms in the brain.

Since the frontiers of scientific discovery are defined as much by the "tools" available for evaluating binocular dysfunctions as by conceptual innovation, I strongly feel that the time is now ripe for Optometry to broaden its scientific horizons and venture into the Cognitive Brain Sciences.

Why?

Because the complexity of human vision is not so much mediated by a pair of eyeballs but by the entire brain. In my opinion, Cognitive Optometry can provide a *complementary* view (no pun) of what role vision plays in the multi-modal and higher order process of human awareness.

Investigating how vision works has several advantages over studying some of our other senses because, as humans, we are highly visual creatures and consider vision to be our predominant sensory modality. Our observations, as vision scientists and also clinicians, are invaluable in providing a more "structured" foundation to anchor endless speculation in the psycho-physical sciences.

Since what we see around us determines how we form our perception of "reality", further exploration of visual "awareness" and what role it plays in the evolution of "consciousness" is crucial in better understanding how neuro-typical patients individually form their own version of "reality".

This concept is just not idle speculation as even a cursory glance at our nervous system and brain reveals that an inordinately large amount of brain tissue (almost 1/3) is dedicated to the processing of visual stimuli. The analysis of images and the prominence of visual competency in correctly interpreting what you see out there enable our visual percepts, as humans, to be so vivid and rich in information.

Intuitively, we note that these sensory modalities mediate quite different sensations and provide insights into seemingly different worlds. Yet, current scientific research utilizing sophisticated imaging techniques reveal that similar neuronal events or circuits in the CNS, in all probability, mediate all three of these subjective sensations or *qualia*.

There is, of course, the underlying assumption that what we usually define as the state of "consciousness" depends on what is going on "inside" the head and not necessarily on the way the animal is "outwardly" behaving i.e. there is an *explicit* correspondence between *any* mental event and its neuronal correlates. In other words, any change in a subjective state must be associated with a change in its neuronal state, induced "naturally" or via neuro-genic drugs.

The psycho-physical discipline of Cognitive Optometry, will seek to recognize the key role played by vision in the complex process of awareness, attention, cognition (understanding) and consciousness (heightened awareness).

This formal study of anomalous mental activity as a visual information-processing problem rests on the implicit assumption that our perceptions, thoughts and social behavior depend on the ability of the visual system to evaluate and collate sensory information correctly. I strongly feel that any inherent uncorrected *refractive* errors in human patients lead to *perceptual* and *conceptual* errors of judgment with induced personality 'issues'.

By "translating" perceptual representations taking place in various regions of the CNS into successful goal-oriented behavior we can help our debilitated patients lead more productive and rewarding lives.

**Dr. Nazir Brelvi**

Allamuchy, NJ

2009

# Section 1

# Introduction

The human brain is a unique amalgamation of ancient brain regions and newer morphologies that have been "tinkered" with by the honing and parsimonious process of natural selection. Evolutionary recent changes wrought within the so-called "association" cortical areas of various lobes have been modified in predictable ways through expansion or reduction of existing ones.

Rapid-fire adaptation to abrupt climatic changes, induced by massive but localized volcanic events or devastation of entire continents by meteorite/comet impacts has played a vital role in shaping our success as apex predators of the natural food chain.

Some prominent evolutionary psychologists believe that the human brain has built into it a so-called "set of rules" that govern our social behavior. Since there an infinite number of habitable environments scattered around the globe, this so-called "tool-kit" has, in their opinion, been used selectively, to successfully emerge as the prime example of natural adaptation.

In the mid to late 1990s, J.H. Kaas, advanced our understanding of cognitive science by challenging the prevailing traditional view of the primate neocortex being composed predominantly of so-called "association" areas. This was a catch-all term assigned to any cortical area that was not sensory or motor and that just happened to border "known" regions.

His work on primate species ranging from *tenrecs* (small insectivores from Madagascar) to *lorises* (small primates found in Southeast Asia) to American *Homo sapiens* over the past few decades, has demonstrated that almost all of the neocortex is sensory and motor in the "higher" mammals.

He found that complex brains evolve not by simply *expanding* association cortex, but by *increasing* the number of sensory and motor 'modules' with massive interconnections between them. A modern analogy would be comparing a two mega-pixel camera to a 100 mega-pixel one, adding several orders of magnitude to processing mental capability.

Theodore Bullock has led the field of comparative neuro-biology for decades. According to him:

"Long before the human species appeared, the pinnacle and greatest achievement of evolution was already the brain - as it has been before mammals appeared, before land vertebrates, before vertebrates.

From this point of view, everything else in the animal world was evolved to maintain and reproduce nervous systems - that is, to mediate behavior, to cause animals to do things. Animals with simple and primitive or no nervous systems have been champions at surviving, reproducing and distributing themselves, but they have limited behavioral repertoires.

The essence of evolution is the production and replication of diversity - more than anything else, diversity in behavior."

Bullock describes well the aims of Comparative Neuroscience and lists them as the three "Rs" of Brain Science: Roots, Rules and Relevance.

Roots refers to the evolutionary history of the brain and behavior. How are brains similar and different in various animal species.

Rules of change are the mechanisms that give rise to changes in the nervous systems over time i.e. are there constraints under which evolving nervous systems develop?

Relevance refers to the general principles of organization and functions that can be extrapolated from a particular animal studied to all animals, including humans.

In other words, even though both the Space Shuttle and the V2 missile are rocket ships, can we, in all fairness, have extrapolated one from the other ?

# Awareness

Complex neural circuitry has evolved in many animals but the common image of animals climbing up some intelligence ladder is misinformed. The commonly held view that 'lower' animals have a few fixed reflexes and that in 'higher' ones the reflexes can be associated with new stimuli (Pavlov's dogs) and the responses can be tagged to rewards (Skinner's rats) is no longer untenable.

Many migratory birds fly thousands of miles at night, maintaining their direction by 'looking' at the stars. If this navigational ability of birds was innate it would soon be obsolete as the precession of the equinoxes, every 25,000 years or so, would have wiped their slates clean. The various birds have actually evolved a special algorithm for learning where the celestial pole is in the night sky. On the other hand, maybe they navigate via special magnetic deposits embedded within their foreheads that use the Earth's magnetic field for steering in 3D space.

Honeybees perform a dance that tells their cohorts the direction and distance of a food source with respect to the sun. The dancer also uses an internal clock to compensate for the sun's trajectory from the time she discovered the food to the time she passes on the information. On cloudy or overcast days, the other bees estimate the direction by using the polarization angle of the diffuse light in the sky.

There are dozens of such examples in the animal world. Many species compute how much time to spend foraging at each patch in order to optimize their rate of return of calories per energy expended. Some birds learn the *emphemeris* function, the path of the sun above the horizon over the course of the day and the year, necessary for navigating by the sun. The barn owl uses microsecond disparities between the arrival times of sound at its ears to swoop down on a rustling field mouse in pitch darkness. Caching species place nuts and seeds in unpredictable hiding places to foil thieves but are able to return months later to retrieve them.

These examples make it amply clear that animal brains are just as specialized and well engineered as their bodies. A brain is a precision instrument that allows a creature to use information to solve the problems presented by its lifestyle. Since organism lifestyles differ species cannot be ranked by IQ or by the percentage of human intelligence they have achieved. They should be viewed instead as stand-alone pinnacles of natural evolution.

Whatever is special about the human mind cannot be just more, better, or more flexible animal intelligence because there is no such thing as generic animal intelligence. Each animal species has evolved its own information processing machinery to resolve its lifestyle issues just as we have evolved ours. I choose to call this neurological process primary or implicit awareness.

Awareness, in my opinion, is a *gestalt* term encompassing attention, social conditioning and innate intelligence. It allows a normal healthy animal to extract information from its surroundings that is relevant to its future wellbeing. It is what makes a cat, a cat. Awareness is tailor-made to serve a particular animal species in its specific ecological niche.

Neuroanatomically speaking, as outlined in my earlier book, attention is also a hierarchical process mediated by the reticular formation (RF) of the brainstem and midbrain in conjunction with various regions of the basal forebrain and prefrontal cortex.

When compared to the vast bulk of the forebrain, the brainstem appears rather small. However, it contains groups of motor and sensory nuclei of widespread modulatory neurotransmitter systems along with numerous ascending and descending tracts that subserve awareness-mediating functions.

Joseph LeDoux of the Center for Neural Science at NYU believes that your brain was assembled during childhood by a combination of genetic and environmental influences. Genes dictated that your brain was a human one and that your synaptic connections, though similar to those of members of your family, were nevertheless distinct.

Then, through experiences with the world, your synaptic connections were adjusted (by natural selection and/or instruction and construction), further distinguishing you from everyone else. The synaptic connections are adjusted by environmentally driven neural activity in specific neural systems.

When these changes occur during early life, they are said to involve developmental plasticity; when they occur later in life, they are considered as learning. However, he feels that the line between these two entities is a fine one and perhaps nonexistent.

# Evolution of the Nervous System

The nervous system is the most complex of all the organ systems in the animal embryo. Billions of nerve cells or neurons develop a highly organized pattern of connections, creating the neuronal network that makes up the functioning brain and the rest of the nervous system.

There are many hundreds of different types of neurons, differing in identity and connectivity even though they may appear quite similar. There are also supporting tissues in the nervous system, known collectively as glia; for example, Schwann cells surround the axons of peripheral neurons whereas oligodendrocytes and astrocytes ensheath the central neurons.

All nervous systems, vertebrate and invertebrate, provide a system of communication through this highly anastomotic network of nerve cells comprising various shapes and sizes. Each neuron connects with its target cell at synapses, sites at which the propagation of an action potential induces neurotransmitter release.

The nervous system can only function properly if the neurons correctly connect to one another and thus a central question surrounding the CNS development is how various neurons connect with such preciseness and specificity.

Specific chemicals acted as intercellular transmitters in the developing zygote's homeostatic system and various chemical messengers housed in its plasma membrane, regulated and controlled its day-to-day activities via these molecular interactions. As the organism became multi-cellular, this 'local' interaction was expanded to reach other component cells, whose activities were also regulated and coordinated for maximum efficiency.

The centralized nervous system with its constituent neurons communicated with each other or with effector organs via chemical agents known as neurotransmitters. Here one neuron alters the activity of the next neuron in a reflex chain by releasing a specific chemical messenger, which diffuses across a very narrow gap (synapse). One important characteristic of chemical mediation that is understandable in terms of membrane receptors is specificity.

A chemical messenger – hormone, neurotransmitter or locally released *paracrine* – influences only certain cells and not others. The explanation is that the membranes or cytoplasm of different cell types, differ in the types of receptors they contain. A single cell usually contains many different receptor types, each capable of combining with only one chemical messenger.

In addition, the receptors themselves are subject to physiological regulation i.e. the number and affinity of the receptors to their specific chemical messenger can be increased or decreased, through a mechanism of feedback control.

How does the messenger-receptor combination elicit the cell's responses?

It usually gets quite complicated and requires the participation of so-called *second messengers*. These are substances generated within the cell or that enter its cytosol via mediated transport. This messenger then triggers the cell's overall response. The same ones may operate in many areas of the body and in different types of cells.

The two best understood second messengers at present are calcium and cyclic AMP (cAMP). Both may sometimes be involved in eliciting the cell's response in the same situation. Also, just because the researchers have isolated calcium and cAMP as *second messengers*, does not mean that these are the only possible ones. Others almost certainly remain to be discovered.

Since the CNS was not designed by some supernatural agency or intrepid engineer, but emerged through the relentless process of natural selection, over immense *eons* of time, neural variability is a fundamental property of the CNS found in any specified animal population. Natural selection, of course, is not the *only* process that modifies organisms or their nervous systems over generations but so far it is seems to be the only known process in nature that appears to *design* them.

During the early stages of evolution, the 'nervous' system was very simple. It 'orientated' the organism within its habitat by mediating movement. In case of the bacterial cell, it propelled the organism towards or away from a source of stimulation, depending on whether it indicated the presence of friend or foe, food or poison. The response was essentially *chemotactic* in nature.

Soon, these membrane receptors, arrayed on the outer surface of the organism, began to involute as the organism became multicellular and multi-layered. They took on the appearance of the synaptic terminals, simple ionic 'gates' initially and evolving into the more complex ones seen today in higher animals.

Next, a more amplified version of a 'truer' nervous system was wrought, with a limited number of interneurons interposed between the afferent and efferent nerve cells. The interneuronal component then began to expand rapidly until it came to be by far the largest part. The interneurons formed smaller 'networks' of increasing complexity, as the housekeeping chores of the organism increased.

At first, the nervous system mediated mainly the stability of the internal environment and position of the body in 3D space. These cells then, with increasing specialization of function, came to be localized at the business *end* of the primitive nervous system. As the organisms became bilaterally symmetrical, the sense organs migrated to the site of the *stoma* or mouth, giving rise to the brain and cranial nerves around it.

According to Stuart Kaufmann, a proponent of Complexity Theory that looks for mathematical principles of order underlying many complex systems in the universe like meteorological phenomena, human societies, crystal formation, galaxies, brain formation etc., such feats of self-organization, order, stability and coherence may be an innate property of some complex systems. "Evolution," he suggests, "may be a marriage of natural selection and self-organization."

His theory does raise some interesting issues. Natural selection presupposes that a "replicator" arose somehow that allowed an organism to react to the vagaries of nature by undergoing mutations for traits that would enhance its coping ability.

Kaufmann asserts that for evolution to occur at all, mutations have to modify a physical body enough to make a difference in its functioning but not so much that it will incapacitate the creature. The process of natural selection would have to work within the constraints of these principles just at it works within the biological ones that define living things.

We tend to view brain evolution as *additive* rather than nonlinear i.e. the complex human brain is simply an ape brain with a few 'new' parts added. In actuality, we should visualize our brain as a unique amalgamation of evolutionary old and newer areas that have been honed and modified to successfully adapt human behavior to the challenges of environmental demands.

Steven Pinker in his recent, euphemistically titled tome: *How the Mind Works*, observes..."the fallacy that intelligence is some exalted ambition of evolution is part of the same fallacy that treats it as a divine essence or wonder tissue or all-compassing mathematical principle. The mind is an organ, a biological gadget. We have our minds because their design attains outcomes whose benefits outweighed the costs in the lives of Plio-Pleistocene African primates."

Evolutionary psychologists do not believe that all behaviors are driven by genetic mechanisms. They think that the brain has built-in adaptations, which are of a more general nature. These adaptations are a set of rules that govern our behavior. Moreover, since there are an infinite number of environments, these rules apply differently in each situation, resulting in an infinite number of behaviors.

In fact, a major assumption in the biological sciences is under siege: the idea that a larger brain with more cells is responsible for the greater computational capacity observed in the *Homo* lineage. Numerous studies and accumulating neurological data show that the human brain's unique abilities are not a function of cell number but the emergence of specialized circuitry.

Split-brain studies, where the corpus callosum is severed leaving behind two isolated 600-650 gm brains rather than just one 1200 gm organ point out that the capacity of the left half (the size of a chimp brain) remains unchanged from its preoperative human level.

Many lines of anatomical and physiological research have found that there are indeed specializations in human brain tissue. The physiological properties of dendritic spines differ from those of other primates. The primary and secondary visual, somatosensory and auditory cortices express varying distributions of specific nerve fibers and the density of certain nerve cells called *chandelier* cells differ between prefrontal and visual cortical regions.

The nonhuman primate and human visual systems also have different organizational properties. When comparing, for example, the anterior commissure between the macaque and humans, the monkeys are still able to process visual information through it when the corpus callosum is severed whereas humans cannot. Lesions to V1 in humans cause blindness whereas monkeys with similar lesions are still capable of residual vision.

Other examples also abound of system-level variations between primates and humans even though both visual systems have virtually identical sensory capacities.

# Development of the Nervous System

The earliest events in embryonic brain development are largely controlled by genes, their products and the local chemical environment in which they exist. When we speak of genetic influences on brain development, we are essentially describing the effects of various proteins and their chemical spin-offs.

In both the brain and body, a small number of genes are able to create complexity, in four different ways. First, the genome encodes structure not as a blueprint but as a process. It specifies a general way or method to build a creature rather than delineate all its details at the outset. Second, the individual genes work in combination and as a team, making the incremental effect of adding a new gene to the genome as exponential rather than linear.

Third, one gene can end up specifying the fate of millions of cells. For example, the so-called master control gene for eye formation, dubbed Pax6, can produce the entire eyeball in all its complexity by simply reusing (switching on/off) the same master instructions over and over again, as often as necessary, in multiple locations.

Finally, a whole bunch of cells can express the same gene, but in a sort of gradient i.e. to different extents. Consider, for example, the set of neuronal connections known as "topographic maps" that run in parallel from the retina to the thalamus, much like the ribbon cables that used to connect a computer to a printer.

To a first approximation, every cell in the retina has the same markers *(Eph* receptors) but in a variant of the gradually diffusing gradients. Individual growth cones use information about how many *Eph* receptors they have - as opposed to their neighbors - to guide themselves to the appropriate target.

Like so many schoolchildren lining up by height along the gymnasium, the axons of individual ganglion cells sort themselves according to their *Eph* levels; moving to the head of the class or to the back depending on which axons have more while the others shift back to make room.

Such gradients allow thousands or even millions of axons to organize themselves in a precise fashion using a tiny number of instructions or genes. The beauty of this system comes in its flexibility. The neuronal axons expanding if there is more space in the mid-brain or bunching up if there is less space than expected.

This auto-regulation gives the building process enough flexibility and also the freedom to adapt to local conditions which would be almost impossible with a rigid system as a blueprint or construction plan.

A few days after fertilization, the intricate process of cell differentiation transforms the *morula* (cluster of embryonic cells into a mulberry-like mass) into an elongated disc. In the third week (after 18 days), the development of the nervous system begins in earnest with the formation of a *neural plate* in the prospective cranial end of the disc.

In a process called *neurulation*, the main body axis forms and the various layers of tissue become well defined. The neural tube and notochord (a stiff rod of cells forming the 'back-bone') emerge and the tube differentiates into the brain and spinal cord. The brain section subdivides into the forebrain, midbrain and hindbrain, which later on give rise to the specialized 'association' areas of the cortex that mediate all voluntary and involuntary activities.

Various so-called *tool-kit* genes for constructing three-dimensional structures are activated, which begin laying down the building blocks of vertebrate brains. The control of individual tool-kit gene action and anatomy is done via numerous *genetic switches* that encode instructions unique to specific species.

These are lengthy sequences of DNA, bound by a variety of proteins, that transform complex sets of inputs into simpler outputs; analogous to a work order in an assembly plant that breaks down the building of a complex machine into simple operations at various stations. Many separate switches may regulate one gene such that it ends up being used numerous times and in different locations.

In the fourth week, three swellings or primary vesicles take shape. The cavity within these pouches is continuous and develops into the ventricular system of the adult brain. The ependymal cells lining the inner wall of the ventricles produce copious amounts of CSF, which now fills it up. The most cranial vesicle forms the forebrain, the middle one gives rise to the midbrain structures and the caudal-most evolves into the hindbrain.

Early in the fourth week, the ventral aspect of the neural tube exhibits shallow, transverse grooves (branchial arches). They represent the first segregation of neurons that later differentiate into the various nuclei of the brainstem. Thus, several cranial nerves and their nuclei are initially laid down in segments corresponding to the *gill arches* of the primordial vertebrate body plan.

The simple layering of the neural tube, with the mantle zone (gray matter) inside and the marginal zone (white matter) outside is retained in the spinal cord. In the cranial part of the neural tube (the future brain), however, major alterations in the mutual positions of these occur forming a layered sheet of gray matter externally just under the Pia.

Also in the fourth week, a shallow furrow, the *sulcus limitans*, at the lower limit of the brainstem, marks the border between the *basal* plates (source of motor neurons) and the *alar* plates (give rise to the sensory neurons). This corresponds to the functional division between the ventral and dorsal *horns* of the spinal cord. In the adult brainstem, the *sulcus limitans* delineates the motor and sensory cranial nerve nuclei. Several neuronal groups in the brainstem migrate from their birthplace in the alar or basal plates.

The cerebral hemispheres begin to form in the fifth week along with the basal ganglia and the internal capsule. As development proceeds, the caudate nucleus and the thalamus come to lie medial to the internal capsule whereas the lentiform nucleus (globus pallidus and putamen) lie laterally.

At the beginning of the eighth week, neuroblasts from the mantle zone migrate into the marginal zone to establish the cerebral cortex. The differentiation from the cortical plate to the six-layered cortex is not complete until after birth and is accomplished by waves of cells migrating toward the cortical surface. Since the deeper layers are laid down first, the neurons destined for the more superficial ones have to "surf" through them guided by the processes of radially oriented glial cells extending from the ependyma to the Pia.

Postnatally, several serial but temporally overlapping processes mark development. There is a continuation of synaptogenesis, which begins prior to birth and in humans occurs at different rates in different brain regions. Synapses in the brain begin to form much before birth, prior to week 27 (counting from conception) in humans, but do not reach peak density for 15 months after birth.

Synaptogenesis is more pronounced early on in the deeper cortical layers and occurs later in the more superficial layers. The neurons are also increasing the size of their dendritic arborizations now, extending their axons and undergoing vigorous myelination.

Synaptogenesis is followed by synaptic pruning, which continues for more than a decade. This pruning allows the CNS to fine-tune neural connectivity. For instance, the ocular dominance columns in the visual cortex initially show a much larger overlap between the projections of the two eyes onto neurons in V1 than after the pruning is complete.

There is compelling evidence that different regions of the human brain reach maturity at different times. For instance, synapses in the superior temporal region of the auditory cortex, reach peak density earlier in postnatal development (age 3 months), than those in the association cortex of the frontal lobes (age 15 months).

Some data also suggest that in humans, the processes of synaptogenesis and synaptic pruning follow different time courses in sensory and motor cortices than the association cortices. Such heterogeneous developmental changes correlate with the developmental time courses observed by Jean Piaget.

The human brain undergoes changes until the late teens, when in most respects it resembles the adult brain. Throughout adulthood, it changes very little in terms of volume, myelination and synaptic density until well into the seventh decade of life, when some reductions in brain volume are seen due to attritional neuronal loss. Yet, recent evidence reveals that new neurons may well be produced throughout adulthood.

Individual neurons can inform others about the strength of the stimulus it receives, such as the intensity of the impinging sound or light waves by varying the frequency and pattern of its firing rate. It seems to be using a frequency code, if you will, to communicate or broadcast its level of stimulation. Even though the firing frequencies vary with different classes of neurons, some are capable of achieving rates in excess of 100 Hz per second.

*Burst* neurons produce trains of action potentials with an intermittent pattern of high frequency discharges, while the single spike ones fire at intervals that are more regular. Others can switch between these two modes depending on which neurotransmitter acts as the triggering agent. Serotonin tends to evoke *plateau* potentials in groups of spinal motoneurons, which play a prominent role in mediating attention or motivation behaviors rather than transmit specific information.

Overall, the code for the information carried by an axon in the human retina is the frequency and pattern of action potentials its cells generate since the strength of action potential is always the same.

# Primary Sensory Receptors: An Overview

Sensory impulses from nearly all regions of the organism's physical body are transmitted to its CNS conveying relevant information about the environment and how the organism is itself reacting to it. The anatomical structures where these sensory impulses originate are called sense organs or receptors.

The main task of such receptor organs is to respond to stimuli of various kinds and intensities. In a process called transduction the receptors help "translate" the stimulus to a language easily understood and "spoken" by the neurons of the nervous system i.e. convert electrical impulses into action potentials.

The stimulus alters the permeability of the receptor membrane, usually by opening the sodium ($Na^+$) channels and thus depolarizing the receptor. The act of depolarization sets up a graded change in the membrane potential in a process similar to the synaptic activation of a neuron. The mechanism behind transduction is not fully understood in its entirety especially when the action potentials lead to a subjective construct or "feeling" in the brain.

Most receptors are built to respond only or preferentially to one kind of stimulus energy, be it mechanical, chemical, thermal or visual. The kind of stimulus to which the receptor responds most easily is called an adequate stimulus. However, most receptors are able to respond to stimuli of other kinds (inadequate stimuli) although the threshold for activation or evoking a response may be slightly higher. We shall explore this built-in capability of receptors in the next section.

Since only humans are able to inform the observer directly of what they feel, animal experiments alone cannot resolve the question of the relationship between receptors and conscious sensations. Not all impulses reaching the CNS from the receptors are consciously perceived for a considerable "editing", if you will, exists at all levels of sensory pathway integration to censor "irrelevant" information and enhance "relevant" ones. This property of the CNS implies that the receptors do not provide the brain with an objective or true representation of its surroundings.

Modern neuroscientists are still unable to explain how action potentials, which are very similar in all nerve fibers, are able to evoke dissimilar conscious sensations in different individuals. They have theorized that the brain has a built-in ability to collate disparate bits and pieces of all relevant sensory information into a coherent and unified percept of the world out there to give it meaning.

A single neuron's membership in any neural network mediating sensory information is probably fluid and may change from moment to moment when it interacts with several such networks in its vicinity. In addition, the neuron's firing cadence may also be modulated by the animal's past perceptual experiences, its internal brain dynamics (whether it is actively or passively sampling its immediate environment) and the animal's expectations for the future.

Brain scientists now believe that the final processing depends on the extensive interconnections found among cortical areas mediating various aspects of sensory information. These specialized areas, found to be clustered in the forehead region of the skull, then interpret which "messages" are being conveyed, assigns a priority status to them and "decide" how the particular individual should respond.

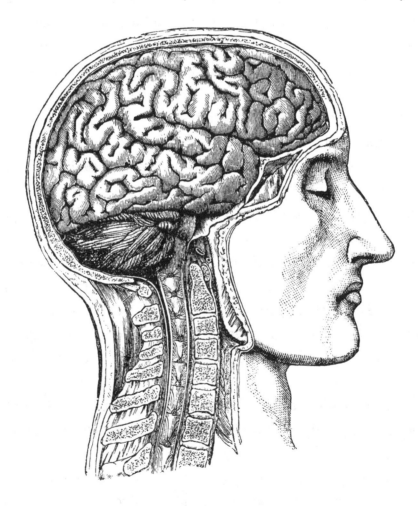

# Sensory Neurons

Sensory fibers from the receptors follow the peripheral nerves toward the CNS. Close to the spinal cord, the sensory fibers are collected in the dorsal roots and enter the cord through these. The sensory fibers of the spinal nerves have their cell bodies in the dorsal root ganglia lying on either side of the spinal cord just as the ones of the cranial nerves lie close to the brainstem.

The dorsal root fibers vary in thickness, from the thickest myelinated ones, with diameters of 2 mm and conduction velocities of 120 m/sec, to the thinnest non-myelinated fibers of less than 1mm, conducting at less than 1m/sec. The thick fibers belonging to the ganglion cells with large cell bodies are often classified as Group A or alpha fibers of varying thicknesses and velocities, down to Group C or slow conduction velocity, un-myelinated ones.

For example, pain sensations are often conducted swiftly to the cerebral cortex via the alpha neurons resulting in a sharp, shooting perception of pain and when the smaller, slower neurons take over, the pain sensation morphs into a slow, burning type of longer duration.

(**Author's Note**: One important point to keep in mind here is that all pathways conducting sensory information from the receptors to the cerebral cortex (except olfactory) are synaptically interrupted in various thalamic nuclei. These nuclei or neuronal groups/clusters are relays in precisely organized, major pathways that reach only certain regions of the cerebral cortex.

They have various functional tasks and they receive fiber connections from the somato-sensory nuclei, the retina, the auditory pathways, the cerebellum, the basal ganglia and others. In addition, the thalamus has a decisive influence on the general level of neuronal activity of the cerebral cortex and thus on the level of consciousness.)

## "Brains"

Why did brains evolve anyway? Why do animals need a brain?

The short answer to this profound query is to process information! Most organisms in order to survive have to make choices every time they decide to move, stop, eat or explore their surroundings. In the unicelled amoeba, *putzing* around in the backyard pond, this may consist of relatively simple tasks of discovering or avoiding. In multicellular animals, these choices rest on how efficiently the information is gathered and converted into crucial decisions. The more you know, the better decisions one can make. The organism's nervous system and brain makes all this possible.

Natural selection cannot directly endow an organism with information about its environment or mental organs to process it but only choose between appropriate genes. However, genes build brains and different genes are capable of building brains that process information in diverse ways. A new field of computer science called genetic algorithms has shown that Darwinian selection can create increasingly intelligent software.

Genetic algorithms are programs that are duplicated to make multiple copies, though with random mutations that makes each one a tiny bit different. All copies then have a go at solving a problem and the ones that do best are allowed to reproduce to furnish the copies for the next round.

But first, parts of each program are randomly mutated again, split in two and the halves exchanged. After many cycles of computation, selection, mutation and reproduction, the surviving programs are found to be better than any a human programmer could have designed.

More apropos of how a mind can evolve, genetic algorithms have been applied to various neural networks participating in a so-called robotic 'thought experiment'. A given network has inputs from simulated sense organs and outputs to simulated motor systems embedded in a virtual environment with scattered 'food' items also made available to numerous other competing networks. The ones that get the most food leave the most copies prior to the next round of mutation and selection.

The 'mutations' are random changes in the connection weights, sometimes followed by 'sexual' recombination (swapping some of their connection weights) between networks. During the early iterations, the automatons wander randomly over the terrain, occasionally bumping into a food source. However, as they 'evolve', they manage to zip from one food source to another with impunity. Indeed, a population of networks that is allowed to evolve innate connection weights often does better than a single neural network left to its own devices.

This is especially true for networks with multiple hidden layers, which complex animals, including some humans, surely possess. It was found that if a certain network could only learn but not evolve, the environmental teaching signal was soon 'diluted' during its propagation back to the hidden layers. It was then only capable of nudging the connection weights up or down by tiny increments.

However, when a network population could evolve without learning, the numerous mutations and recombinations it underwent during the simulation were able to reprogram the hidden layers directly. In fact, this occurred to such an extent that the network's innate connections were fine-tuned to an optimum configuration. The geeks found that its innate structure was selected for.

Evolution and learning can also go on simultaneously with the innate structure evolving in an animal that also learns. A population of networks can be equipped with a generic learning algorithm and can be allowed to evolve the innate parts that the network designer would ordinarily have built in by guesswork, tradition or trial-n-error.

The innate specs include how many units there are, how they are connected, what the initial connection weights are and how much the weights should be nudged up or down on each learning episode. Simulated evolution gives the networks a big head start in their learning careers.

Interestingly, James Baldwin, a psychologist from the turn of the century, had proposed that learning could also guide evolution in precisely this way, creating an illusion of Lamarckian evolution without there actually being one. Called the Baldwin Effect, it is now thought to play a key role in the evolution of brains. Contrary to traditional social science assumptions, learning is not some pinnacle of evolution attained only recently by human beings.

All but the simplest of animals are capable of learning. That is why mentally uncomplicated creatures like fruit flies and sea slugs have been convenient subjects for neuroscientists in search of the neural correlates of learning.

If the ability to learn was in place in an early ancestor of multicellular animals, it could have guided the evolution of nervous systems toward their specialized circuits even when they are so intricate that natural selection could not have found them on its own.

# Basic Architecture of "Brains"

When any primate brain is viewed in a container filled with formalin in any neuro-lab, one notices that it is encapsulated in three coverings called the dura, the pia and the arachnoid. The dura is the toughest and surrounds the brain, all around and in between. The pia is very thin and membranous, containing most of the brain's extensive blood supply.

The arachnoid is sandwiched between these two layers and is filled in with the cerebro-spinal fluid (CSF), which apart from other functions adds buoyancy to the brain tissue. This effectively protects the brain from being jostled around and reduces its overall weight from a robust 3 pounds in humans to a mere 150 grams.

After the dura is peeled away, it reveals the pinkish-gray walnut-like surface split in the middle by a huge fissure. The raised rounded sections are called gyri and the deeper fissures or creases underlying the purplish blood vessels are known as sulci. A quick dissection through the hemispheres exposes some internal structures organized into white and gray matter.

The gray matter forms a continuous sheet overlying a homogenous mass of white matter. The gray matter contains cell bodies of neurons and glia whereas the white matter contains the fatty myelin sheathed axons of these neurons.

The pia consists of an outer epipial layer and an inner membranous one called the intima pia. This layer invaginates, where the blood vessels enter and leave the CNS, forming a perivascular space. It derives its nutrition from the CSF and underlying neural tissue. The epipial layer consists of a meshwork of collagenous fiber bundles continuous with the arachnoid trabeculae. The blood vessels of the spinal cord lie within the epipial layer. The cerebral vessels lie on the surface of the Intima pia within the subarachnoid space.

In the region of the conus medullaris, the epipial tissue forms a covering of the filum terminale, fine processes of deeply located fibrous astrocytes. The CSF fills up the subarachnoid space between the delicate, non-vascular arachnoid and the epipial layer. Several invaginations along the deeper sulci of the brain result in a pooling of the CSF forming discrete 'cisternae'.

The ambiens or superior cistern surrounds the posterior, superior and lateral surfaces of the midbrain. It is of special significance as it contains the Great Vein of Galen, the posterior cerebral, superior cerebellar arteries and the Pineal gland, adjacent to the III ventricle. CSF formed in the lateral and 3rd ventricles passes via the cerebral aqueduct into the 4th ventricle.

From there it flows out of the median and two lateral apertures into the cerebello-medullary cistern, flows out into the contiguous subarachnoid spaces of the brain and spinal cord. The bulk of the CSF returns to the venous system via the arachnoid granulations and villi. The rate of exit is pressure dependent and one-way only.

In regions adjacent to the Superior Sagittal Sinus (SSS), multiple tufted prolongations protrude through the meningeal layer of the Dura. These granulations are variable in number and location, each consisting of numerous arachnoid villi surrounded by venous lacunae. These passive one-way valves have membranes, which are readily permeable, allowing even plasma proteins and serum albumin to pass through.

The rate of CSF formation in humans is approx. 600-700 ml/day. Since the total volume of the ventricles and subarachnoid spaces is only about 140 ml in volume, the daily turnover seems to be appreciable. The CSF has a nutritive function and that it serves to remove the waste products of neuronal metabolism.

The characteristic distributions of ions and non-electrolytes in the CSF and plasma imply that the CSF cannot be a simple filtrate or dialysate of the blood plasma. Current evidence supports the theory that it is a secretory product involving active transport mechanisms with expenditure of cellular energy.

Recent studies indicate extensive plexi of serotonin axons in the supra and sub-ependymal systems lining the walls of the ventricles and the arachnoid sheaths draping the major cerebral blood vessels originating in the Raphe nuclei of the Pons.

These may be major modifiers of the local CSF composition and cerebral blood-flow through vascular regulation. In addition, the presence of such biogenic amines in the CSF could help mediate the release of hormones along the Hypothalamic-Pituitary axis.

The human brain represents a unique amalgamation of evolutionary older and newer areas that were modified in predictable ways. Some that underwent an expansion, some reduction, whereas others ended up wired in novel ways as the organism adapted to fresher environmental demands.

Evolutionary psychologists believe that the brain has inbuilt adaptations (a set of general rules that govern behavior) that functionally contribute to the organism's propagation. In their view, the brain does not perform any significant metabolic, mechanical or chemical service for the animal but serves a purely informational, computational or regulatory function.

A detailed mapping of these abilities is an indispensable first step in the neuroscience research enterprise. Two prominent methods are the PET (positron emission tomography) and functional MRI (fMRI). Both detect subtle changes in metabolism or blood flow in the brain while the subject is actively engaged in cognitive tasks. As such, they enable the researchers to identify brain regions that are activated during these tasks and to test hypotheses about functional brain anatomy.

# White Matter

The so-called white matter of the brain appears milky-white due to the fat-laden myelin surrounding the axons. If we were to simply look at the brain in gross dissections, we would not know that structures such as the white matter were neural elements rather than supportive tissue.

To make this inference, neuro-anatomists must probe for finer detail with high-powered microscopes as in Histology, the study of tissue structure through numerous staining/microscopic techniques. A primary concern is to identify the patterns of connectivity in the nervous system in order to lay out the neural "highways" that allow information to get from one place in the brain to another.

This problem is made complex by the fact that neurons are not wired together in a simple serial circuit. A single cortical neuron is innervated by many neurons and the axons from these input neurons can originate in widely distributed regions of the cortex. There is tremendous "convergence" in the nervous system as well as "divergence", where a single neuron can project to multiple target neurons in different parts of the cortex.

Neighboring and distant connections between two cortical regions are referred to as *cortico-cortical* connections, following the convention that the first term identifies the source and the second term, the target. For example, thalamo-cortical would mean the fiber bundle begins at the thalamus and ends within the cortex, whereas, *cortico-thalamic* would imply the reverse.

Most axons are short projections from neighboring cortical cells. Others can be quite long, traveling through so-called fiber tracts, descending below the cortical sheath into sub-cortical, brainstem or even spinal cord nuclei.

# Myelination of the CNS & Cortical Plasticity

Myelin is a fatty covering or insulating sheath wrapped around the axons which run from a neuronal cell body to its neighbor like an outstretched, elongated finger. Modern investigation has revealed that nerve impulses race down axons on the order of 100 times faster when they are coated with myelin. This myelin is wrapped up to 150 times around the axon and is manufactured in sheets by two types of glial cells.

The oligodendrocyte, an octopus-shaped glial cell does the wrapping and the insulation thickness corresponds to the axon's diameter. The other glial cell called as the Schwann cell are able to detect a protein called neuregulin, whose concentration determines how thick the myelin sheathing of a particular axon needs to be. Interestingly, many people who suffer bipolar disorder or schizophrenia have a defect in the gene that regulates neuregulin production.

The wrapping occurs at different ages. Myelin is prevalent in only a few brain regions at birth, expands in spurts and is not fully laid until age 25 or 30 in various places. Myelination generally proceeds in a wave from the back of the head forward as we grow into adulthood. The frontal lobes are the last places to be myelinated. These regions of the cortex are responsible for higher-level reasoning, planning and judgement - skills that only come with experience.

Critical periods occur at different times and vary in duration for different systems and behaviors. In some cases, they may begin with a genetically determined proliferation of synapses in a particular neuronal network. For example, in human children, intense training and effort for walking on two legs is kicked off around age one followed by an explosive development of vocabulary between ages of two and three (especially if the toddlers tumbles down a long flight of stairs).

(**Author's Note:** The term critical period is applied to the development of a particular function of the nervous system when that subsystem is maximally plastic, i.e. its capacity for structural and functional adaptation is complete. Once these sensitive periods have passed, the CNS displays a marked decrease of plasticity.)

Although language may develop even if training starts many years later (as observed in epileptic children), it may not achieve full developmental potential as it was not instituted early enough. A similar logic applies in the display of amblyopic symptoms in children with small angle strabismus that have "missed" the critical period of optimal visual acuity from soon after birth to two years of age. Various aspects of vision have different critical periods and at the cellular level, neurons in the different strata of the visual cortex develop their characteristic properties at different times.

It is a commonly held notion that we lose a large number of neurons as we age. This is partially based on the fact that the brain is on average 8% lighter at age 80 than at 25. However, how much of this weight loss is actually caused by cellular attrition or a mild dehydration of cellular tissue as the cell bodies undergo shrinkage is debatable. Numerous studies on normal aging and diffusely distributed neuronal loss cannot readily be correlated with altered behavior or reduced mental functions.

Another popular notion that any damage to the CNS in the adult leads to irreversible damage; that neurons do not regenerate damaged connections nor does the brain replace lost neurons has come under increasing debate as recent studies suggest some degree of neurogenesis in adult humans. After all, most adults are capable of learning throughout their lives. This learning is a sure proof of brain plasticity for it involves changes in synaptic weights between neurons in the brain's circuitry, as observed in long-term potentiation.

Interestingly, brain-imaging studies show that elderly and young people have different patterns of cortical activation during cognitive tasks, even when performance is equal. Specifically, processes that are strongly lateralized in the young are more evenly divided between the two hemispheres in the elderly. Such observations strongly suggest that plastic changes do occur in the aging brain presumably re-routing neuronal traffic to counteract the detrimental effects of neuron attrition etc. by streamlining flow.

This cortical plasticity is not limited to the somatosensory or tactile modality but also observed in the auditory and visual systems. The visual cortex has a detailed topographic map of the visual world called the retinotopic map. (See brain maps and visual space for details).

Interestingly, amputation of a limb induces a reorganization of the cortex subserving its functions, leading to bizarre patterns of perception in that individual. Researchers discovered, in a dramatic example of plasticity in humans, that the region previously *coding* for the missing limb, became *functionally* innervated by the adjoining cortex. However, since this cortex was mediating the face region, the patient reported feeling sensations in the *phantom* limb when stroked on the *face* with a Q-tip.

The term *functional* plasticity is used because the effects may not be caused by a physical reorganization in cortical neuronal circuitry. Rather, in the normal case, the receptive fields of contiguous neurons overlap mediating the so-called phantom limb phenomenon.

Biologists have also noted that domesticated animals had smaller brains than their "wild" cousins did. This neural 'hypertrophy' seemed to occur quite early because animals born in the wild and later tamed ended up with brains of the same size as the wild ones. The difference is not genetic but environmental as the individuals of the first generation born in captivity have smaller brains also.

Researchers attribute this phenomenon to the differences for behaviors required to dwell in the tumultuous wild as opposed to snoozing on the family couch. Lab experiments confirm the profuse dendritic arborizations found in the cerebral cortex of rats raised in a simulated natural environment with ample space and access to toys than in rats restricted to standard cubbies.

Another example evaluated the density of optic fibers leading from the retina to the visual cortex. Initially, neurons in the cortex are influenced with equal strength from each eye. However, soon after birth, the neurons in the cortex segregate into groups with one eye providing a larger portion of neural input than the other does. Termed ocular dominance, this implies active competition for the available synaptic sites at V1, the primary visual cortex neurons. The eye with precise optics usually wins out and is highly influential in formulating visual percepts beyond V1.

The previous examples of brain size and environmental influence on dendritic arborizations strongly suggest that synaptogenesis is closely linked to adaptation and learning. The decisive factor was not the amount of "awake" time but how this time was utilized i.e. the learning of specific coping skills and new behavior patterns rather than zoning out being a couch potato.

It has become increasingly clear in recent years that alterations of brain structures with advancing age are not uniformly distributed but concentrated in specific regions of the brain. Psychological testing reveals that not all capacities of the human brain are significantly reduced by normal aging. An investigation of more than 1600 individuals asked to repeat a list of 20 words just presented showed that the performance of the oldest group (88 years) overlapped that of the youngest (25 years) by 50%.

Thus, even memory is not as strongly correlated with age as is often assumed. Many researchers describe plasticity as a harbinger of an expansion of human potential in which the powers of the brain will be harnessed to revolutionize child-rearing, education, various therapies and aging.

# The Corpus Callosum

The two cerebral cortices are interconnected by the largest fiber system in the brain named the corpus callosum. In humans, this bundle of white matter includes more than 200 *million* axons that tend to link so-called homotopic regions of the cortex. For example, the left primary visual cortex will connect with it's right counterpart etc.

Callosal fibers also project to heterotopic areas of the cortex. These tend to mirror the ones found within a hemispere i.e. visual areas may interconnect with auditory or vestibular nuclei providing a neural network basis for multi-modal perception (synesthesia). Even though scientists have basically "mapped-out" most of the callosal projections, their functional role still remains elusive.

For example, some researchers point out that in the visual association cortex, receptive fields can span both visual fields. Communication across the callosum enables information from both visual fields to contribute to the activity of these cells. In this view, callosal interconnections could play a key role in *synchronizing* oscillatory activity in adjacent cortical neurons as an object passes through the various contiguous receptive fields. In other words, the callosal fibers facilitate processing of *directional* information in the 3D visual space by pooling together diverse inputs in *real* time.

As with the cerebral hemispheres, researchers have investigated functional correlates of anatomical differences in the corpus callosum. Usually investigators measure gross aspects like the cross-sectional area of the callosum. Variations here are linked to gender, handedness, mental retardation, autism and schizophrenia, among numerous other neurological impairments.

Major callosum sub-divisions are organized into functional zones whose posterior regions are more concerned with visual information and whose anterior regions mediate auditory and tactile information.

The anterior part of the callosum is involved in higher-order transfer of semantic information. Interestingly, this information tends to "describe" certain aspects of the stimulus rather than tell what the actual stimulus is.

For example, so-called "split-brain" patients report through the "speaking" hemisphere only the items flashed to the right half of the screen and deny seeing left-field stimuli or recognizing objects presented to the left hand. Furthermore, the left hand correctly retrieves objects presented in the left visual field of which the patient verbally denied having any knowledge.

# The evolving Cerebral Cortex (CC)

The comprehensive exploration of the cortex had to await the advent of modern microscopy, biochemical stains and various dyes that selectively bind to diverse cellular elements. With this enhanced ability to target specific molecular constituents of neurons, the cortex has been catalogued and mapped quite extensively.

The human cerebral cortex is very intricately organized and considered one of the most densely populated tissues in the body. It has about 30 billion neurons in total, where each cubic mm houses 100,000 neurons. Each of these neurons has upwards of 10,000 connections or synapses. No one has really made an exact count of the different types of neurons in the brain, but it could be in the neighborhood of about fifty.

One general characteristic of its operation is the astonishing variety and specificity of the actions it performs. Sensory systems handle an almost infinite variety of images, scenes, sounds, etc., which react to them in detail with remarkable accuracy. The actual body of a single neuron may measure a few tens of a micron in diameter but its axon could extend from a few mms to even a meter in length. Flanking each of these neurons is a glial (glue like) cell, which supports and nourishes it.

In order to better understand, consciousness as a process and an emergent property of the CNS, the reader needs to have a usable knowledge of its architecture, its anatomical organization and the remarkable dynamics it generates. The adult human brain weighs about 3 pounds and is among the most complicated objects in the universe. It is certainly one of the most remarkable biological structures to have emerged during the evolutionary process containing upwards of a 100 billion neurons.

Confronted with a structure as complex as the cortex, most scientists divide it into increasingly smaller and smaller sections in order to understand it better. A deeply held belief among biologists is that structure and function may be intimately related i.e. if any increase in the cellular density is detected or changes are noticed in the degree of myelination or certain neurological enzymes mysteriously emerge in the CSF, then it is very likely that some functional border has been broached.

Until the mid-1970's, it was widely believed that structure and function are intimately related i.e. differences in structure are reflected in differences in function and vice versa. If the cellular packing density increases or the degree of myelination changes or some enzyme makes a debut, then it is very likely that a functional border has been crossed.

Do the deep folds or grooves discernable on the outside of the cerebral hemispheres demarcate disparate regions subserving different functions? Or, are these just folds resulting from shoving a large pancake-like cortex sheet into a cramped cranial cavity? What is the exact relationship among all these areas? Do the interconnections reveal anything novel about its large-scale architecture? Are these areas randomly bunched together to fit inside the skull or is there some underlying hierarchy?

A key principle of neuronal architecture is that nearby neurons encode similar information. This widespread feature of the cortex and other nervous tissues economizes on total wiring length. Spatial clustering also shows up in different ways.

The various lobes of the CC play a diverse functional role in neural processing. Major identifiable systems can be localized within each lobe, but as we shall see some of the major systems like hearing or sight are much more widespread and highly interactive.

While all of the neocortex behind the central sulcus deals with sensory input and perception, the concern of the frontal lobes, the larger expanse of cortical matter lying forward to it is action oriented. Motor, premotor, prefrontal and anterior cingulate cortices all belong to the frontal lobes. Cognitive brain systems rely on both cortical and sub-cortical components to act as modulators of their main activities.

The frontal lobe plays a key role in the planning and execution of movements. Its two main subdivisions known as the motor contains massive clusters of neurons whose axons extend to the spinal cord and brainstem. The other, called the prefrontal, mediates the higher aspects of motor control and tasks that require the integration of information over longer time periods. Its also maintains inter-connectivity with structures of the limbic lobe.

Areas of the parietal lobe receive inputs from the somato-sensory relays of the thalamus that represent information about touch, pain, temperature sense, and limb proprioception.

(**Author's Note:** The volume of neocortex that cannot be categorized as sensory or motor has traditionally been termed the association cortex. These areas receive collateral inputs from one or more modalities, which serve to provide "contextual or orientational" information in order to correct any inaccuracies in performing the task at hand.)

Numerous animal studies with microelectrodes reveal segregated cortical neighborhoods whose neurons are specialized to carry out different jobs. For example, neurons in one occipito-temporal region are especially sensitive to the hue of a certain object, whereas the ones in a region of the posterior parietal cortex mediate complex eye movements. These observations are borne out very well in human patients that present with focal deficits secondary to trauma or stroke episodes.

Cortical areas in the back of the brain, known to scientists as the visual cortex, seem to follow a 'loose' hierarchy comprising of at least a dozen levels, each subordinate to the other. When a group of neurons within one of these regions receives a strong input from a 'lower' layer, it undergoes a rapid 'weighted-average' evaluation and gets 'bumped-up' to the next 'higher' layer in the hierarchy, analogous to an irate patient's complaint being kicked-up to the entry-level employee's manager.

If the incoming information is kind of 'sketchy' or incomplete and not amenable to an unambiguous interpretation, then that particular network will 'fill-in' the deficits, arriving at a 'best guess' solution. This type of 'filling-in' occurs throughout the cortex. If the irate patient is still unhappy, the complaint rises further up the 'chain-of-command'.

The brain scientists are puzzled as to why this hierarchy exists. One reason may be that such an architecture, permits 'higher' cortical regions to 'track' correlations among neurons in 'lower' areas. In other words, keeping with the 'corporate' analogy, the entry-level 'lower' personnel carry on with their 'housekeeping' chores, while the 'higher' corporate executives progressively make 'broader' and 'more important' decisions.

Modern-day cognitive neurobiologists no longer describe the various nuclei and regions of the brain but instead concentrate on three major topological systems that mediate its global functioning. These are:

(1) The thalamo-cortical system, which consists of a dense meshwork of reciprocal connections between the thalamus and numerous cortical regions,

(2) a long, polysynaptic loop system organized in a set of parallel, unidirectional chains, which link the cortex to the Cerebellum, Basal ganglia and the Hippocampus, and

(3) An extensive but diffuse set of connections emanating from the Locus Coeruleus and Raphe nuclei of the brainstem and projecting out like a fan into the cerebral cortex.

The thalamus and the cortex evolved in close relationship to each other. Except for the sense of smell, all sensory modalities relay through the thalamus and on to the cortex. Neurons in the numerous thalamic nuclei fall into two broad classes: excitatory projection ones that convey impulses to the cortex and local, inhibitory interneurons.

The projection neurons are further categorized into core and matrix ones. Core cells aggregate in clusters and target precisely delineated recipient zones in the intermediate cortical layers of different regions. Matrix projection neurons reach in a more diffuse manner into the superficial layers of several contiguous areas of the cortex. They help disperse and synchronize the activity of literally hundreds of cells.

While all of the neocortex situated behind the central sulcus (the deep groove that partitions off the front part of each hemisphere from the back), mediates sensory input and perception, the large expanse of neocortex lying forward of it is geared for 'action'. As organisms evolve, the complexity of their actions increases and their goals extend in space and time, coming to depend less on instinctual drive and more on prior experience, insight and reasoning.

This necessitates planning, decision-making in uncertain environments, cognitive control, recall and differential diagnoses. The prefrontal cortex mediates all these so-called high-level executive issues. It integrates information from all sensory and motor modalities.

The various components of the brain function not only as individual processors of discrete signals but also 'collate' their 'graded' responses and interact with each other to produce an integrated, unified conscious response. Although the overall pattern of connections of a given brain region is describable in general terms, the microscopic variability in any human is enormous, giving it a unique signature.

Together with the morphological peculiarities of the brain and its neural connections with the body's sensory receptors, the animal is provided with a large set of constraints whose role in fostering species-specific perceptual categorization or adaptive learning cannot be underestimated.

The way in which patterns of activity in the brain change with time produces a widespread synchronization of many different functionally specialized areas, helping to integrate the sensory and motor processes. This integration ultimately gives rise to perceptual categorization, the ability to discriminate an object or event from a background for adaptive purposes.

# The Prefrontal Cortex (PFC)

The PFC is especially well developed in humans, is present in other primates, rudimentary in nonprimate mammals and doesn't even exist in other creatures. In humans, the prefrontal cortex constitutes a massive network that links the brain's motor, perceptual and limbic regions. It occupies nearly half of the entire frontal lobe.

The PFC is further organized into a lateral, medial and ventral portions, each of which mediates specific and distinct executive functions, working memory and planning of activities. While other mammals have medial and ventral PFCs, primates alone appear to have lateral PFC. Thus, one of the major regions involved in working memory in primates clearly does not exist in other animals.

The fact that the cognitive capacities of these creatures do not compare with those of primates suggests that the unique features of primate cognition came with the development of the lateral PFC and its integration with existing networks involving the medial and ventral areas.

There are extensive projections to the PFC from almost all regions of the parietal and temporal cortices and even some from the prestriate regions of the occipital cortex. Subcortical structures including the basal ganglia, cerebellum and various brainstem nuclei project indirectly to the PFC via the thalamus.

Indeed, almost all cortical and subcortical areas influence the PFC either directly or within a few synapses. The PFC also sends reciprocal connections to most areas that project to it including the premotor and motor ones. The PFC has many projections to the contralateral hemisphere; not only to the homologous prefrontal areas via the corpus callosum but also bilateral projections to the premotor and subcortical region.

From these neuroanatomical considerations, we can safely assume that the PFC is in an excellent position to coordinate cognitive processing across wide regions of the CNS.

# Subjective Sensory Experience

Sensory impulses from nearly all parts of the body are relayed to the CNS bringing information about how the ambient conditions in the environment are affecting the various tissues and organs. The task of the receptors is to respond to various stimuli and communicate their presence to the nervous system.

Most receptors are built to respond only or preferentially to just one kind of stimulus energy: mechanical, chemical, thermal, light etc. and this property then enables their classification into mechano-receptors, chemo-receptors, pain & temperature receptors or light receptors.

The kind of *conscious* sensory experience a receptor evokes is often uncertain as animals are not able to communicate how they "feel" whereas we as sentient human beings, can. However, not all impulses reaching the CNS from the receptors are consciously perceived. For example, enteroceptive (from internal organs) signals are usually processed at a sub-conscious level.

For signals from all kinds of receptors, however, a considerable selection and suppression of signals (editing) takes place at all levels of the sensory pathways in the spinal cord and brain. Thus, "irrelevant" (not related to task at hand) information is deleted and "relevant" impulses undergo amplification.

This "coded" information is then routed to various "modules" in the sensory cortex, which "sort" and prioritize the incoming information so a decision or appropriate response can be obtained in "real" time. The integration of sensory information from different modalities may not necessarily be due to convergence of all relevant information onto a single cell group.

As we shall see later on in the book, we now believe the final processing of sensory information to inherently depend on extensive inter-connections among cortical regions dealing with very specific aspects of object location, relevance and recognition.

# Multi-modal Integration

Where and how do our various senses get fused in the brain?

The brain receives information from the sensory organs via different channels. Only by combining this information – sensory integration – can we gain an overall unitary image of our surroundings.

Two basic mechanisms are conceivable. Either the senses function separately and our brain combines their inputs into a coherent whole during the final stages of processing

Or, the senses work together from the beginning, complementing and enhancing each other at a very early stage.

Around the 1970s, as psychologists were investigating sensory integration from a perception point of view, scientists coming from more classical biological fields such as neurophysiology, began to investigate the neuronal basis of how the brain combines sensory information from disparate sensory modalities.

Whereas many of these researchers concentrated on specific senses like the visual and auditory pathways in isolation, a small minority began to study the neural network mediating or exhibiting multi-sensory properties.

Only recently, helped in part by advances in brain imaging techniques, such as the PET, fMRI and CT scanners, have scientists begun to realize that our different senses do not function as discretely as was previously thought.

A series of imaging studies have disclosed a complex network of brain regions that are activated most strongly when various sensory data fuse. This integration occurs early on during the processing of neuronal stimuli. Even brain centers that specialize in a given sense use information from other sensory channels. It has long been known that the so-called 'association' areas in the parietal and frontal lobes of the cerebral cortex process information streaming in through numerous sensory channels.

However, regions that up to now have been thought to mediate one specific sensory modality have recently been demonstrated to possess a broader neurological 'footprint'.

For example, certain regions of a superordinate auditory region – the secondary auditory cortex – have now been shown to process visual and tactile stimuli, lending credence to the highly myopic patient's quip about how if they put on their eyeglasses they would hear you better.

In 2005, Christoph Kayser of the Max Planck Institute for Biological Cybernetics in Tubingen and his team performed a series of high-resolution magnetic resonance scans on the auditory cortex of the macaque monkey. They found that a kind of sensory integration seems to occur in the *posterior* section of the auditory cortex that provides the monkey with information on the *directionality* of a particular sound in its environment.

This region of the auditory cortex seems to fuse incoming sensory information and registers *spatial* as well as auditory information. Just like humans, the macaque auditory cortex comprises of numerous sub-units only a few millimeters thick that encircle the primary auditory cortex like a belt. Other researchers have also documented the existence of a neural mechanism by which simultaneously applied non-auditory (tactile) stimulation leads to an enhancement of neuronal activity in the auditory cortex.

Visually speaking, the superior colliculus (SC) is part of the tectum, located in the midbrain, superior to the brainstem and below the thalamus. It contains seven layers of alternating white and gray matter, of which the superficial contain topographic maps of the visual field while the deeper layers contain overlapping spatial maps of the visual, auditory and somatosensory modalities (Affifi & Bergman, 2005).

The SC receives afferents directly from the retina, as well as from various regions of the cortex (primarily the occipital lobe), the spinal cord and the inferior colliculus. It sends efferents to the spinal cord, cerebellum, thalamus and occipital lobe via the LGN. The SC contains a high proportion of multi-sensory neurons and plays a vital role in the motor control of orientation behaviors of the eyes, ears and head.

Receptive fields from somato-sensory, visual and auditory modalities converge in the deeper layers to form a two-dimensional, multisensory map of the external world. Here objects straight ahead are represented caudally and peripheral objects, rostrally. Similarly, locations in the superior field are represented medially while the inferior field is represented laterally (Stein & Meredith, 1993).

However, in contrast to simple convergence, the SC integrates information to create an output that differs from the sum of its inputs. Following a phenomenon called the 'spatial rule', neurons are excited if stimuli from multiple modalities fall on the same or adjacent receptive fields, but are inhibited if the stimuli fall on disparate fields (Giard & Peronnet, 1999).

Excited neurons may then proceed to innervate various muscles and neural structures to orient an individual's behavior and attention toward the stimulus. Neurons in the SC also adhere to the 'temporal rule' in which stimulation must occur within close temporal proximity to excite neurons. However, due to the varying processing time between modalities and the relatively slower speed of sound to light, it has been found that the neurons may be optically excited when stimulated some time apart (Miller & D' Esposito, 2005).

Clearly, many regions of the brain are actively engaged in collating and integrating information from different sensory modalities. Researchers are also finding that a much smaller section of the cortex mediates each individual sense than previously thought.

# Object Recognition

Object recognition depends primarily on the analysis of the form of a visual stimulus, although cues such as color, texture and motion certainly contribute to normal perception. To account for shape-based recognition, two things need to be considered. The first has to do with shape encoding i.e. how is the shape internally represented. What salient features enable us to recognize differences between a triangle and a square or a monkey and a person?

The second center on how shape is processed when the perceiver's viewing position is rarely constant. We can recognize shapes from a wide array of positions and orientations since our recognition system is not hampered by scale changes in the retinal image as we move closer to or further away from an object.

A central debate in object recognition has to do with defining the frame of reference where recognition occurs. Two general approaches have been proposed: view-dependent and view-invariant recognition.

In view-dependent theories, perception is assumed to depend on recognizing an object from a specific viewpoint. They posit that we have a plethora of specific representations of varied objects in our memory bank. The viewer simply comes up with the closest appropriate match by visually rotating the object in mental space. The time needed to decide if two objects are similar or different increases as the viewpoints diverge.

In the view-invariant approach, object recognition does not occur by simply analyzing the stimulus information. Sensory input defines its basic properties and recognition may depend on an inferential process based on a few salient features.

Sensory information can change from one viewpoint to another. Nonetheless, the external world has certain basic architectural details that the visual system exploits during perception. The various "theories" of object recognition have emphasized the significance of such invariant properties with which we are able to infer our surroundings.

# Maturation of Subcortical Visual Circuits

The foveal region in the human retina is immature at birth, whereas the peripheral retina is more developed. Hence, newborn vision is driven predominantly by peripheral inputs. In a similar way, the optic nerve is not myelinated completely in the newborn but it proceeds at a rapid pace, reaching adult-like patterns within two years. The LGN, acting as the main relay between the retina and cortex also experiences rapid growth in the first six months, almost doubling in volume.

The primary visual cortex matures in stages, the deeper layers which project to the superior colliculi (SC) developing earlier than the more superficial ones. The SC is the main subcortical target of the retinal ganglion cells and mediates eye movements (oculomotor and saccadic). It is the first subcortical circuit to be myelinated in the visual array and primarily drives the infant visuomotor behavior.

By one month of age, the infant is able to fixate for longer periods. Coincident with this behavior is the development of projections from the striate cortex (V1) to the subcortical structures that inhibit activity in the SC i.e. automatic overt orienting is replaced by selective fixation.

By two months, infants develop smooth pursuits and begin to show normal orienting to novel stimuli presented in the visual field (VF). The infants also begin to attend to the internal features of complex stimuli as their macular region matures leading to enhanced visual acuity. The pattern in smooth pursuit may result from the coincident development and maturation of V1 projections to the middle temporal (MT) motion areas, which pathway is critical for mediating them.

Between three and six months, infants are capable of anticipatory eye movements. This is possible via maturation of projection pathways from the upper layers of V1 to the frontal eye fields responsible for voluntary eye movements. Now their visual acuity reaches normalcy and these infants are able to integrate the two eyes into a binocular percept.

# Development of "Face" Recognition

Facial feature recognition is an exquisitely developed skill in humans that has its origins in the first days of life. Newborns seem to like looking at human faces or face-like stimuli more than they like looking at abstract patterns. Infants just a few weeks old can distinguish their mother's face from other women's faces but since they have such poor visual acuity, they primarily rely on global aspects (smile, hairline etc.) of her facial features.

After 3-4 months of age when their acuity becomes more normal, they are better able to distinguish among various faces and even have a toothy smile for other family members. However, event-related potentials recorded in response to faces suggest that face processing does not fully mature until puberty. Thus although precocial skills are seen in newborns, face processing requires a good deal of experience before it is fully developed.

This early stage of "facial awareness" may be mediated by subcortical pathways, which in turn help to shape the further development of the association cortices to which they send neural projections. Over time, the constant fixation on numerous facial profiles around them and correlated neural activity associated with this behavior "wire-up" the higher-order regions of the brain to process faces in more adult-like patterns. Based on available developmental data, the early precortical stage is relatively short-lived in comparison to the later more protracted adult-like phase of face recognition.

# Language? Acquisition

Humans are not born with the ability to understand or speak a particular language; they must learn it through exposure and practice. However, humans and some higher primates do possess an innate ability to acquire or learn to speak and understand a language. This may sound semantic but there is an essential difference. For example, let us take human bipedality.

Though our parents guided us in learning to walk as infants, even without any instructions we would eventually stand up and begin to walk. It can be categorized as a normal characteristic of Homo sapiens that occurs at a certain stage of human development. Walking has dedicated neural structures and circuits that are common to all humans around the globe. It is an innate ability whereas acquiring a language is not.

That said, most humans also are able to acquire any language during childhood they are exposed to. Some neural structures that mediate the ability to acquire spoken languages like semantics and grammar appear to be subserved by nonidentical neural systems regardless of the speaker's native tongue.

Humans seem to have an innate knowledge of linguistics that enables them to employ syntactic structures of language. They are thus able to develop complex forms of linguistic representation that is part of the human brain's specialization for language.

# Brain Plasticity & Critical Periods

It is obvious from the dramatic cellular events observed during embryogenesis and the formation of the nervous system that it is quite plastic or malleable during the course of its development. This plasticity stands in stark contrast to the relative rigidity seen in the adult brain. The ability of the mature brain to process sensory information critically depends on the system being used during a specific period of time in early postnatal development.

During postnatal development, the brain as a whole undergoes numerous structural changes heralding a normal period of most rapid neural proliferation. In some notable instances, such as the orientation columns of the visual cortex, factors concerned with normal activity via afferent inputs fine-tune the functional connectivity within the cortex.

However, this postnatal plasticity is limited: the cells are not free to migrate to new areas or to make large changes in long-distance connectivity. In contrast, local cortical connectivity can be affected during so-called sensitive periods, when extrinsic influences can alter brain organization.

(Author's Note: The term critical period is applied to the development of a particular function of the nervous system when that subsystem is maximally plastic, i.e. its capacity for structural and functional adaptation is complete. Once these sensitive periods have passed, the CNS displays a marked decrease of plasticity.)

Critical periods occur at different times and vary in duration for different systems and behaviors. In some cases, they may begin with a genetically determined proliferation of synapses in a particular neuronal network. For example, in human children, intense training and effort for walking on two legs is kicked off around age one followed by an explosive development of vocabulary between ages of two and three (especially if the toddlers tumbles down a long flight of stairs).

Although language may develop even if training starts many years later (as observed in epileptic children), it may not achieve full developmental potential as it was not instituted early enough. A similar logic applies in the display of amblyopic symptoms in children with small angle strabismus that have "missed" the critical period of optimal visual acuity from soon after birth to two years of age. Various aspects of vision have different critical periods and at the cellular level, neurons in the different strata of the visual cortex develop their characteristic properties at different times.

It is a commonly held notion that we lose a large number of neurons as we age. This is partially based on the fact that the brain is on average 8% lighter at age 80 than at 25. However, how much of this weight loss is actually caused by cellular attrition or a mild dehydration of cellular tissue as the cell bodies undergo shrinkage is debatable. Numerous studies on normal aging and diffusely distributed neuronal loss cannot readily be correlated with altered behavior or reduced mental functions.

Another popular notion that any damage to the CNS in the adult leads to irreversible damage; that neurons do not regenerate damaged connections nor does the brain replace lost neurons has come under increasing debate as recent studies suggest some degree of neurogenesis in adult humans. After all, most adults are capable of learning throughout their lives. This learning is a sure proof of brain plasticity for it involves changes in synaptic weights between neurons in the brain's circuitry, as observed in long-term potentiation.

Interestingly, brain-imaging studies show that elderly and young people have different patterns of cortical activation during cognitive tasks, even when performance is equal. Specifically, processes that are strongly lateralized in the young are more evenly divided between the two hemispheres in the elderly. Such observations strongly suggest that plastic changes do occur in the aging brain presumably re-routing neuronal traffic to counteract the detrimental effects of neuron attrition etc. by streamlining flow.

This cortical plasticity is not limited to the somatosensory or tactile modality but also observed in the auditory and visual systems. The visual cortex has a detailed topographic map of the visual world called the retinotopic map. (See brain maps and visual space for details).

Interestingly, amputation of a limb induces a reorganization of the cortex subserving its functions, leading to bizarre patterns of perception in that individual. Researchers discovered, in a dramatic example of plasticity in humans, that the region previously coding for the missing limb, became functionally innervated by the adjoining cortex. However, since this cortex was mediating the face region, the patient reported feeling sensations in the phantom limb when stroked on the face with a Q-tip.

The term functional plasticity is used because the effects may not be caused by a physical reorganization in cortical neuronal circuitry. Rather, in the normal case, the receptive fields of contiguous neurons overlap mediating the so-called phantom limb phenomenon.

Biologists have also noted that domesticated animals had smaller brains than their "wild" cousins did. This neural 'hypertrophy' seemed to occur quite early because animals born in the wild and later tamed ended up with brains of the same size as the wild ones. The difference is not genetic but environmental as the individuals of the first generation born in captivity have smaller brains also.

Researchers attribute this phenomenon to the differences for behaviors required to dwell in the tumultuous wild as opposed to snoozing on the family couch. Lab experiments confirm the profuse dendritic arborizations found in the cerebral cortex of rats raised in a simulated natural environment with ample space and access to toys than in rats restricted to standard cubbies.

Another example evaluated the density of optic fibers leading from the retina to the visual cortex. Initially, neurons in the cortex are influenced with equal strength from each eye. However, soon after birth, the neurons in the cortex segregate into groups with one eye providing a larger portion of neural input than the other does. Termed ocular dominance, this implies active competition for the available synaptic sites at V1, the primary visual cortex neurons. The eye with precise optics usually wins out and is highly influential in formulating visual percepts beyond V1.

The previous examples of brain size and environmental influence on dendritic arborizations strongly suggest that synaptogenesis is closely linked to adaptation and learning. The decisive factor was not the amount of "awake" time but how this time was utilized i.e. the learning of specific coping skills and new behavior patterns rather than zoning out being a couch potato.

It has become increasingly clear in recent years that alterations of brain structures with advancing age are not uniformly distributed but concentrated in specific regions of the brain. Psychological testing reveals that not all capacities of the human brain are significantly reduced by normal aging. An investigation of more than 1600 individuals asked to repeat a list of 20 words just presented showed that the performance of the oldest group (88 years) overlapped that of the youngest (25 years) by 50%.

Thus, even memory is not as strongly correlated with age as is often assumed.

# Cognitive Psychology

The discipline of Cognitive Psychology emerged in the early 1960s in response to the limitations of behaviorism. While attempting to retain the experimental rigor of behaviorism, cognitive psychologists focused on mental processes that were more complex and closer to the domain of psychoanalysis.

Much like the psychoanalysts before them, the new cognitive psychologists were not satisfied with simply describing motor responses elicited by motor stimuli. Rather, they were interested in investigating the mechanisms in the brain that intervene between a stimulus and response – mechanisms that convert a sensory stimulus into action.

The cognitive psychologists set up behavioral experiments that allowed them to infer how sensory information from the eyes and ears is transformed in the brain into images, words or actions. The thinking was apparently driven by two underlying assumptions: the first was based on the Kantian notion that the brain is born with *a priori* knowledge "independent of experience".

That idea was later advanced by the European school of Gestalt psychologists, the forerunners, together with psychoanalysis, of modern cognitive psychology. The Gestalt psychologists argued that our coherent perceptions are the end result of the brain's built-in ability to derive meaning from the inherent properties of the world around us; limited only by features detectable by our multi-modal sensory systems.

In their view, the reason the 'brain' can derive meaning from say a limited analysis of a visual scene is that the visual system does not simply record a scene passively as a video-camera. Rather, human perception is creative: the visual system transforms the two-dimensional patterns of light on the retina into a logically coherent and stable interpretation of a three-dimensional sensory world.

Built into neural pathways of the brain are complex rules of "guessing". These rules allow the brain to extract information from relatively impoverished patterns of incoming neural signals and convert it into a meaningful image.

Cognitive psychologists demonstrated this ability with studies of optical illusions i.e. misreadings of visual information by the brain because it expects or is pre-programmed to "see" a certain image. The brain's expectations are built into the anatomical and functional organization of the visual pathways, derived in part from experiences (explicit) but in large part from the innate wiring (implicit) for vision.

To appreciate these evolved perceptual skills, it is useful to compare the computational abilities of the brain with those of artificial computational or information-processing devices, referred to in the section on Embodied Cognition.

The second assumption developed by cognitive psychologists was that the brain achieves these analytic triumphs by developing an internal representation, if you will, of the external world – a cognitive map – and then using it to generate a meaningful image of what is out there to see, hear and smell. This so-called cognitive map is then combined with information about past events and modulated by attention.

Finally, the sensory representations are used to organize and orchestrate purposeful actions.

The idea of a cognitive map proved to be an important advance in the study of behavior and brought cognitive psychology and psychoanalysis closer together. It also provided a view of mind that was much broader and more interesting than that of the behaviorists. However, this concept was not without problems. The biggest problem was the fact that the internal representations inferred by cognitive psychologists were only sophisticated guesses; they could not be examined directly and thus were not readily accessible to objective analysis.

In order to "see" the internal representations – to peer into the black box of mind – cognitive psychologists had to join forces with biologists. Fortunately, at the same time that cognitive psychology was emerging in the 1960s, the biology of higher brain function was maturing.

During the 1970s and 1980s, behaviorists and cognitive psychologists began collaborating with brain scientists. As a result, neural science, the biological science concerned with brain processes, began to merge with behaviorist and cognitive psychology, the sciences concerned with mental processes.

The synthesis that emerged from these interactions gave rise to the field of cognitive neural science, which focused on the biology of internal representations and drew heavily on two lines of inquiry: the electro-physiological study of how sensory information is represented in the brains of animals and the imaging of sensory and other complex internal representations in the brains of intact, behaving human subjects.

# The Brain's 'Tracking' System

A so-called "cognitive map" of the outside world was isolated recently in the rat hippocampus by researchers at the Norwegian University of Science and Technology. Previously, the hippocampus was known as the seat for memory, by most researchers since the 1970s, acting as an encoding mechanism for recording the timeline of its lifespan.

In 1971, John O'Keefe and Jonathan Dostrovsky had found neurons in the hippocampus that displayed evoked potentials signaling place-specificity. Dubbed 'place' cells, these neurons seemed to organize the various aspects of experience within a framework of "locations and contexts" in which events tend to occur.

In other words, these hippocampal neurons would fire whenever a rat occupied a specific location in the environment but would remain quiet when the rat was foraging elsewhere. Similar findings have been reported subsequently in numerous other animal species including humans.

This remarkable discovery led researchers to propose that the hippocampus was the neural locus of a 'cognitive map' of the environment. They argued that hippocampal 'place' cells "organize the various aspects of experience within the framework of the locations and contexts in which events occur". This contextual framework then encodes *relationships* among the different aspects of the event in a way that allows later retrieval from the subject's memory.

This view has been hotly debated for years. However, a consensus is emerging that the hippocampus does somehow provide a spatial context vital to the 'laying down' of episodic memory. In other words, when one remembers a past event, one remembers not only the people, objects and other discrete components of the event but also the spatio-temporal aspects of the situation. This allows the subject to differentiate among similar events with similar components.

Despite intensive study, however, the precise mechanisms by which the hippocampus creates this contextual representation of memory have eluded scientists. A primary confounding factor being a paucity of data regarding collateral neural networks that feed the hippocampus with pertinent information.

Early work suggested that the entorhinal cortex, an area of cortex adjoining the hippocampus, might encode such spatial information. However, a team of Norwegian scientists made an amazing discovery. They uncovered a system of 'grid' cells in the medial entorhinal cortex that, much unlike a 'place' cell, which typically fires when a rat occupies a specific location in ambient space, each 'grid' cell will fire when the rat is in any one of many locations arranged in a uniform *hexagonal* grid.

Rats (and presumably humans) have thousands of specialized neurons labeled 'grid' cells that seem to track the animal's location in the environment like an autonomous GPS system imprinted into the cortex. Each grid cell projects a virtual latticework of equilateral triangles across its surroundings and fires whenever the rat reaches one of the corners.

These grid cells, conclude the researchers, are likely to be key components of a brain mechanism that constantly updates the rat's sense of its location, even in the *absence* of external sensory input. This could serve as an outstanding model system for gaining further insights in how the brain constructs 'cognitive representations' of the world "out there" that are not explicitly tied to any sensory stimulation.

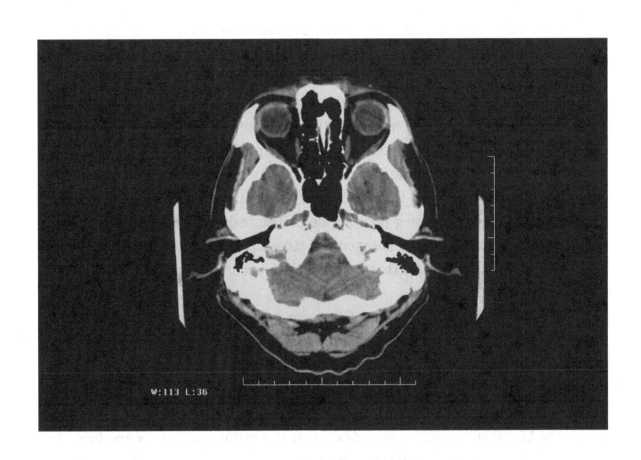

# Section 2

# Sensory Receptive Fields

Operationally, the receptive field of a neuron is defined as the region in the visual field in which an appropriate stimulus modulates the photo-receptor cell's response. In the lab, electro-physiologists frequently connect the amplified output of the implanted micro-electrode monitoring the active neuron to a loudspeaker. This enables them to actually hear the differences in crackling sounds emitted by the firing neuron (evoked potential) during the experiment.

Depending on the profile of this auditory cue, numerous types of receptor cells have been discovered in the visual cortex, mediating the quality and intensity of the light stimulus impinging upon its field. The receptive field of most retinal ganglion cells possess a spatial antagonistic structure (on-center /off periphery or off-center/on periphery etc.) that constructs an accurate sketch or line drawing of the visual field.

This information flows along the 6 cm long optic nerve of each eye and merges with those from the other at the optic chiasm. Here the right/left visual field input undergoes a hemi-field fragmentation (see figure) and heads for the thalamus and LGN. It is the best known of several thalamic nuclei that process visual information.

The LGN is strategically placed between the retina and the cortex. Incoming retinal information is switched over onto a geniculate relay neuron that sends this data onward to the primary visual cortex. The receptive field of the projection cell is nearly identical to that of its input fibers, to such an extent that it is erroneously assumed, by most researchers that no significant transformation of the retinal input occurs here.

The forward projection from the LGN to the V1 layer of the cortex is much smaller than the V1 feedback channels. This suggests that the cortex is selectively editing (enhancing or suppressing) the retinal input passing through the LGN. The visual input is collated into discrete regions and analyzed for shape/

size and numerous other constancies previously embedded into the master visual program.

Collateral interneurons then dispatch appropriate "copies" to the limbic system and surrounding nuclei for various emotional/visual memory tagging. These multiple pathways, with over a million fibers, carry more than 10 million bits of visual information per second. That is faster than your standard landline or online DSL hookup!

By the time, the retinal input enters the nine principal areas of the visual cortex, it has already been broken down into individual pixels, if you will, each reflecting minute changes in the visual percept of the binocular field out there.

Besides a large projection to the superior colliculus, discussed next, numerous minor pathways access an assorted collection of vision mediating nuclei that evaluate blinking, gaze direction, pupillary size, ambient light and other regulatory functions. Since these nuclei do not have a visual map, they do not play a conscious role in cognition.

About 100,000 ganglion cell axons run from the retina to the superior colliculus (SC) straddling the midbrain. The SC is the most important visual processing center in fish, amphibians, turtles and other reptiles. In primates, much of its function has been usurped by the cortex.

It does however coordinate the fine tracking or pursuit movements of the eyeballs via the extra-ocular muscles and mediate the eye/head movements, orienting responses and vestibulo-ocular reflex (from the inner ear). This helps dampen the constant jiggling of the visual field as you look around and walk at the same time.

Some retinal ganglion cells send their axons directly into the inferior pulvinar nucleus. The rest of the pulvinar receives its visual input via the superior colliculus while three of the predominantly visual nuclei project to a wide region of the neocortex including the posterior parietal, the inferior temporal, the prefrontal and orbito-frontal areas.

While the eyes are essential for normal forms of seeing most of the neurons that underlie visual cognition, do not encode eye-of-origin information in an explicit way. They are only concerned with the Big Picture aspects of your vision and the diverse neuronal structures in the thalamus and cortex read out the optic nerve signals and generate a stable, homogenous and compelling view of the world out there.

# Brain Maps & Visual Space

Neural pathways conduct information to and from sensory and motor neurons in the body and the cerebral cortex. Signals conveying sensations such as touch and pain arrive in the foremost part of the parietal lobe at the somatosensory cortex or the sensory *homunculus*, a Latin term meaning "little man". Here the signals coalesce to form a person's body image as the nerve cells represent sensory receptors for different parts of the body in an organized map.

Every animal body's sensory surface and external worlds are represented in such detailed cortical maps, which provide a point-for-point representation of neurons responding to visual stimulation. There is a map for the hands, face, torso and each extremity. The higher the resolution a sensory surface has, the more neurons there are to represent that area. For instance, the neurons that respond to fingertips are greater in number and more densely packed than the ones responding to the back of the hand making them more sensitive to touch. This phenomenon is known as the *cortical magnification factor*.

The body maps also seem to follow a logical sequence where the neurons coding for the index finger are situated next to the middle finger, next to the ring finger etc. This is known as somatotopy and such cortical maps are called somatotopic maps. The maps are present in all adult animals and humans and appear to be the basis for the ordering of our perceptions. They reflect the receptive field properties of cortical neurons.

Many cells in motor areas represent both sensory and motor information. This is true not only for neurons in the parietal cortex but also for neurons in motor areas of the frontal cortex and in sub-cortical structures such as the basal ganglia, cerebellum and superior colliculi (mediate eye movements).

These receptive fields are dynamic and depend on real-time feedback of proprioceptive signals forecasting how various structures of the physical body are distributed in three-dimensional space based on an internal frame of reference. This enables or facilitates dealing with the many real-world tasks encountered in daily life, with their often conflicting demands. As the popular whine goes: "nuthin' is ever easy".

# Independent or Convergent Pathways

Thus far, we have emphasized that the visual system contains multiple pathways; each specialized to extract specific information. Yet the outputs from these pathways are designed to complement each other. This paradox has resulted in a lively debate concerning the inherent independence of such pathways.

Advocates favoring a segregationist viewpoint focus on anatomical, physiological and behavioral data. They indicate that the processing of features such as motion and color involve separable mechanisms from the first synapse in the CNS all the way to the extra-striate cortex. To these theorists, neurological dissociations are the crowning piece of evidence.

On the other side are those who emphasize the lack of perfect segregation. Cells in the M and P-blob pathways exhibit some orientation selectivity. Color sensitivity is not the exclusive domain of the P-blob pathway but also found in the P-interblob ones. Lesion studies of animals have also brought into question the notion that the processing of features like depth, color and orientation depends solely on a single pathway.

For these reasons, the term concurrent processing was coined by David Van Essen, which emphasizes the analytical characteristics of visual perception. As noted from the foregoing discussion on PET imaging, there is strong evidence that the visual system does not analyze the input en masse but breaks it up into its various constituents (line drawings or sketches) and then builds up the gestalt or global percept of the visual scene.

# Neuronal Gestalts?

In neuro-speak, a gestalt, by definition, is a highly variable aggregation of neurons that is temporarily recruited around a triggering epicenter. This idea was recently introduced by Susan Greenfield of Oxford University, London to serve as a *model* for an *intermediate* level of neuronal activity, lying between a single neuron and the entire cerebral cortex.

A single brain cell or neuron in isolation does not contain any ethereal element of consciousness. Furthermore, we have seen that once we venture beyond animals such as the sea slug or lobster, we do not have the one-to-one matching of physiology with 'psychology'.

For example, in the so-called 'lower' animals it is fairly easy to predict responses based on physical stimulation. But this ability gets progressively "fuzzy" as we go higher up in the animal hierarchy and into the province of sentient mammals.

Therefore, in order to overcome this obstacle it becomes necessary to conceptualize how aggregations of neurons go about their business and end up 'translating' this physical stimulus into a valid response or behavior. Neuronal gestalts are also needed to investigate whether or not they can function in both time and space.

These *transient* aggregations of neurons form, operate and re-form at any instant in multiple regions of the brain, such that at any moment, one particular group somehow "generates" what we commonly refer to as awareness or primary consciousness. They do not have a rigid or fixed anatomical location and thus are not necessarily restricted to just one brain region.

It is further thought that the size and disposition of such neuronal assemblies are strongly influenced by the subject's state of arousal i.e. on the activity level of the brainstem RAS.

The cerebral cortex seems to be organized into so-called barrels or modules where neurons in each layer reach up and down, at right angles to the brain surface, to connect in vertical columns. As the cortex becomes increasingly dominant in animals that are higher up in the hierarchy, the critical factor that changes is the potential complexity of connections among these cells.

A close analogy is how the potential for communication is increased dramatically by adding more telephone lines as the office gets larger. This is an important point because in the cortex, the functional groups of neurons need not be locked into a perpetual dialogue and the relevant connections need not be functionally fixed or predetermined. Anyone can or is able to connect with any other person in the office equipped with a phone.

Same with these neuronal gestalts. They are essentially *transient* groupings of neurons, coming into existence as and when needed in keeping with the activity level of the office.

# Neuro-matrix

In the late 1980s, psychologist Ronald Melzack of McGill University and his colleagues theorized that illusory body parts arise at least in part from neural activity within the brain. In their view the brain not only detects sensory signals from the body but also generates its own neural pattern or signature that provides it with a so-called "body image" in its intact state. This signature inscribes the psyche with a sense of the body's configuration and borders that an individual can call his own.

This self body image (neuromatrix) persists even after the removal of a body part, creating the mistaken perception that the ablated organ is still present and attached to the body. The basic structure of our "neuromatrix" may be present at birth and most likely inscribed in our genome as the "phantom limb" phenomenon has been reported in infants born without these extremities.

The brain's somatosensory cortex contains a map of various body regions based on the tactile information it receives from the body via a sensory pathway that traverses the thalamus. Yet another neural pathway transmits information from the body to the limbic system, which governs emotions such as those associated with phantom limbs. Thus, after the loss of a body part, neural activity in this network system may result in the perception of an actual limb where there is none.

Single-cell recording studies have provided physiologists with a powerful tool to map out the visual areas in the macaque monkey brain and characterize the functional properties of the neurons within them. This work has also provided strong evidence that different visual areas are specialized to represent distinct attributes of the visual scene. Inspired by these results, researchers have employed neuro-imaging techniques to ask whether a similar architecture can be discerned in the human brain.

Semir Zeki of University College in London used PET (positron emission tomography) to verify that different visual areas are activated when subjects are processing color or motion information. Subtractive logic factored out the difference in the activation modes in the control and experimental situations. The results of the two studies provided clear evidence that the two tasks activated distinct brain regions.

Although the spatial resolution of PET is coarse, these areas were determined to be in front of the striate (V1) and prestriate cortex (V2). In contrast, after the appropriate subtraction in the motion experiment, the residual foci were bilateral but near the junction of the temporal, parietal and occipital cortices. These foci were more superior and much more lateral than the ones for color. They were labeled as human V4 for the color foci and V5 for the motion task (most researchers now refer to this area as human MT even though it is not in the temporal lobe of the human brain).

The activation maps in this PET study are rather crude. More recently, sophisticated functional magnetic resonance imaging (fMRI) techniques have been applied to study the organization of the human brain and visual cortex in particular. In these studies, a stimulus is systematically moved across the visual field. The blood oxygen level dependent (BOLD) response for areas representing the superior quadrant will tend to increase at a different time than the response for areas representing the inferior quadrant, allowing the entire field to be continuously tracked.

In order to compare areas that respond to foveal stimulation and those that respond to peripheral stimulation, a dilating and contracting ring stimulus is used. By combining these different stimuli, researchers can measure the cortical representation of the contralateral visual field.

Because of the convoluted nature of the human visual cortex, the results from such an experiment would be indecipherable if we were to plot the data on the anatomical maps found in a brain atlas. To avoid this problem, a flat representation is constructed. High-resolution anatomical MRI scans are obtained and computer algorithms are employed to transform the folded, cortical surface into a two-dimensional map by tracing the gray matter.

The activation signals from the fMRI study are then plotted on the flattened map, with color-coding used to indicate areas that were activated at similar times. Boundaries between visual areas are defined by reversals in the retinotopic representation within a quadrant and the foveal regions are well identified by a string of asterisks.

A number of maps encode object position (location) using implicit representations that depend on the actuator concerned. Thus, the eye movement system contains a different representation of visual space than the brain region encoding visually guided reach movements.

Some neurons do explicitly encode object location in the brain. So-called place cells have been isolated from the rodent hippocampus, which fire maximally when the animal is physically within a particular region in its environment. These neurons remain silent outside this restricted area.

Functional imaging of the human brain has revealed object-specific zones in the cortex. The sight of objects selectively activates the ventral temporal (VT) cortex, including the fusiform gyrus and the lateral occipital region. Most researchers agree that the sight of human faces preferentially activates the so-called fusiform face area (FFA) in the fusiform gyrus. Induced lesions in this neighborhood are often associated with an inability to recognize faces.

Currently, a debate rages between localists who assign one chunk of the ventral stream to the dedicated analysis of faces, another one to body parts, a third sector to houses and spatial scenes. Holists argue that object re-awareness is more widely distributed in patches of overlapping activity throughout the brain.

# Neuro-Magic: "Doors" of Perception

Magicians have been testing and exploiting the limits of human cognition, awareness and attention for hundreds of years. Magic tricks often work by covert misdirection, drawing the spectator's attention away from the secret "method" that makes a trick work.

Some neuroscientists are scrutinizing such magic tricks to gain an insight or probe aspects of our awareness, which may not be necessarily grounded in sensory reality. Current brain imaging scans have revealed that some regions of the brain are especially active during certain kinds of magic tricks.

Magicians are, first and foremost, artists of attention and awareness. They manipulate the focus and intensity of human attention, controlling at any given instant, what the audience is or is not fully aware of. They do so in part by employing bewildering combinations of visual illusions (such as after-images), optical illusions (smoke & mirrors), special effects (explosions, fake gunshots, precisely timed lighting controls etc.), sleight of hand, secret devices and mechanical artifacts (gimmicks).

However, the most versatile instrument in their bag of tricks may be the ability to create cognitive illusions. Like visual illusions, cognitive illusions mask the perception of physical reality. Yet, unlike visual illusions, cognitive illusions are not sensory in nature. Rather, they involve high-level CNS functions such as attention, memory and causal inference.

With all these tools at their disposal, well-practiced magicians make it virtually impossible to follow the physics of what is actually happening - leaving the impression that the only logical explanation for the unfolding events, is magic.

The investigators found greater activation in the anterior cingulate cortex among the subjects who were watching magic tricks than among the controls. The finding suggests that this brain region may be important for interpreting causal relationships. These findings also begin to suggest the powerful ability of magical techniques in manipulating attention and awareness while studying the physiology of the brain.

If neuroscientists learn to use the methods of magic with the same skill as professional magicians, they too should be able to control awareness precisely and in real time. If they correlate the content of awareness with the functioning of neurons, they will have the means to explore some of the mysteries of human consciousness itself.

# Ensemble Coding or Re-awareness

The finding that neurons of the IT region of the temporal lobe selectively respond to complex stimuli agrees with the hierarchical theories of object perception. According to the Ensemble theory, V1 neurons code elementary features such as line orientation and color of the various objects in the visual field. Their outputs are then combined to form detectors sensitive to higher-order features such as corners, edges or intersections, an idea consistent with the findings of Hubel and Wiesel.

The process of reconstructing the outer scene progresses through each successive stage coding for combinations that are more complex until the collated information arrives at the IT. The neurons here are activated by specific shapes like hands or faces. This type of neuron has been called a *gnostic* unit, referring to the idea that they can signal the presence of a *known* stimulus - an object, place, animal or person that has been encountered by the observer in the past.

These so-called *gnostic* cells are representational elements that are infrequently active and so carry a lot of information when they do fire. Their firing corresponds to a complex feature i.e. mother's face/profile/voice/ smell etc. in the sensory input. Thus, a small number of them can represent an entire scene and they can classify or categorize specific areas of the field in sensible ways.

With this population or ensemble hypothesis, re-awareness is not due to one unit but to collective activation. The ensemble theories readily account for why we tend to confuse one visually similar object or person with another. Losing some of these Gnostic cells might also degrade our ability to recognize an object.

Interestingly, it was discovered that the selectivity of these neurons is usually relative rather than absolute. They prefer certain stimuli to others but can also be "faked" into firing by ambiguous signals that may be visually similar.

Are human or animal faces special? Is there such a thing as a "Grandma" cell?

Some scientists assert that face perception may not be mediated by the same processing mechanisms as the ones for object re-awareness. One argument in favor of this hypothesis is based on primate and human evolution. The reasoning goes something like this: When we meet other individuals, we usually scrutinize their faces rather than the rest of the body, a behavior that is not newly acquired by humans but which has a long-standing history.

The tendency to focus on faces is found across most cultures around the globe and the study of facial expressions provide the most salient clues to the person's moods or emotional state.

Two decades of research confirmed that cells in two distinct regions of the temporal lobe are preferentially activated by faces; one in the superior temporal sulcus and the other in the inferior temporal gyrus. Numerous imaging studies have also confirmed engagement of the fusiform gyrus during a range of face perception tasks. Indeed, this region has come to be referred to as the fusiform face area (FFA) in literature.

However, even in the right hemisphere of the brain, the FFA does not appear to be exclusively activated by faces. Tasks requiring judgments about dogs or trees also produce activation in the FFA over baseline but to a lesser extent than actual faces. Thus, even though face perception appears to utilize distinct physical processing systems, we must keep in mind the problems with *single* dissociations as in the so-called "grandma" cell.

As the numerous studies point out, patients who have a selective disorder in face perception but have little problem with recognizing objects, does not imply a specialized processor mechanism for just faces. Perhaps the tests that assess face perception are more sensitive to the effects of brain damage than the ones evaluating objects.

When we consider the kind of tasks typically involved in assessing faces as opposed to objects, the patient has to decide between two faces (same category) versus a broad range of objects (different categories). Thus, face perception tasks involve within-category discrimination whereas object perception tasks involve between-category ones.

The goals-oriented aspects of perception that we use vision to guide our movements, to manipulate tools or to recognize faces underscore the selective aspects of awareness. Humans, and possibly some of the great apes are not passive processors of information. We can select from the dazzling array of neural impulses impinging upon our senses at any one time. Depending on our goals, the relative importance of various sources of information constantly changes.

# Section 3

# What is Vision?

The process of natural selection did not design the visual system to entertain us with pretty vistas and gorgeous colors but to deliver a sense of appreciation for the tangible forms and objects present in the real world. The selective advantage is obvious: creatures that are adept in finding new food sources, avoiding predators and refrain from dangerous pursuits will survive long enough to procreate and flourish.

Vision, as a form of remote viewing enables organisms to assess the environment from a distance and initiate appropriate responses well before actual physical contact becomes inevitable. Evolution of the visual system dawned when the diffuse sunlight penetrating the shallow reefs succeeded in sparking a sensory reaction from a surface receptor embedded in the skin as a simple function of the energy content of photons.

Over millions of years, the differing wavelengths of the electro-magnetic radiation (EMR) spectrum began to invoke visual sensations of various hues, some visible and others invisible. Stimuli for visual sensations came in two flavors: threshold and subliminal. At threshold, most primates discern wavelengths ranging from 400 to 750 nm (nanometers) resulting in the rainbow hues of perception. As we shall see later on, birds, turtles and fish along with numerous species of insects can also see in the ultraviolet and infrared regions of the EMR.

The subliminal include any non-light events that yield any sensation of seeing. Such stimuli generally produce unformed sensations called "phosphenes or photopsias". The pressure phosphene appears as a patch of contrasting light and dark when mechanical force is applied to the exterior of the eyeball.

Electrical phosphenes are bright purplish swirls observed with the passage of a weak electric current through the eyeball. Numerous orbiting astronauts have reported scintillating phosphenes, presumably due to cosmic rays, during the course of their space walks and extra-vehicular excursions.

Vision provides the viewer with a realistic depiction of the world out there that is relevant and useful for its day-to-day well-being. Many animals have two eyes and whenever they face forward so that their fields overlap, the brain has to fuse two disparate images of the world into a single unified whole. This unified image was termed Cyclopean after Homer's mythological creature that possessed a single eye in the middle of its forehead.

The inherent problem with such a cyclopean image is that there is no direct way to overlay the views of the two eyes. Objects closer than the point of fixation tend to wander outward toward the temples while the closer ones squeeze in towards the nose. However, through a simple process of triangulation, natural selection was able to figure out how far the object of regard was when the two monocular views were fused. This binocular percept of depth that evolved in the higher predatory animals is known as stereoscopic vision.

Bats, killer whales and dolphins navigate via echolocation i.e. vision mediated through ultrasound. As the wavelengths get larger and consequently slower (reduced cycles/second), vision overlaps with sound. Just as humans and primates are highly visual creatures, each animal species has evolved to manipulate their sensory systems to optimize information gathering within the context of their own unique ecological niche.

The propagation of EMR (electromagnetic radiation) may be thought of as waves lapping in the ocean. Primate vision, then, is best likened to a bunch of surfers only excited by the sight of high rollers ranging between 400 and 750 cm (13 to 25 feet) tall. Any waves above or below this range are either unexciting or too dangerous to navigate. However, such radiation may also be considered as a fine spray that coalesces into "solid" waves of different heights.

Nuclear phenomena such as Einstein's photoelectric effect and the detection of low levels of light by the retina are best explained by the quantum detection theory, a quantum being the smallest unit of detectable energy. The energy content of a single quantum is proportional to its frequency: it is greatest at the short wavelengths and lowest at the longer ones.

The frequency of an electro-magnetic (EM) wave is constant regardless of the medium in which it is traveling. However, the velocity of light varies with the medium of traverse, slower through water or glass than through the vacuum or air. The ratio of the velocity of light in vacuum (or air) to the velocity of light in an optical medium is known as the index of refraction of that substance. As a convenience, the index of refraction of air is rounded off to the value of one.

Since the velocity of light in air is always greater than that in a denser medium, the index of refraction of glass or water is taken to be higher than one. Note that, although the frequency of light is a constant, its wavelength varies with the refractive index of the traversed substance. Thus, light traveling at 510 nm in air will have a shorter wavelength at the retina level (after passing through the denser aqueous and vitreous) of only 382 nm. Does this mean that the green color in air is perceived as blue by the retina?

No, because in actuality, each optical medium has a different refractive index for each frequency of light and the retinal photoreceptors are calibrated accordingly. The central retina is designed for optimal sensitivity to wavelengths between 530 & 550 nm (amber or greenish-yellow). The blue or shorter wavelengths are focused forward of the retinal plane and the reds behind the actual retina. Both these colors are perceived as somewhat blurry and give rise to the phenomenon of chromatic aberration.

Now, although the visible portion of the EM spectrum is characterized physically by their wavelength and energy content, these do not in any way specify the subjective sensation of vision they induce. In fact, the efficiency of light in producing a color percept tends to change with its wavelength not proportional to its energy content. For example, fewer watts of green light are needed to produce a given sensation of brightness than any other spectral color.

This property of human vision, called trichromacy, arises because our retina uses only three types of light-absorbing pigments to mediate color. Although trichromacy is common among primates, it is not universal in the animal kingdom. Almost all non-primate mammals are dichromats, with color vision based on just two kinds of visual pigments. A few nocturnal mammals have only one pigment.

As mentioned above and also in the Color Vision section of my first book, some birds, fish and reptiles have four visual pigments capable of detecting UV light, invisible to humans. The short wavelength (S) pigment absorbs light maximally at 430 nm, the medium wavelength (M) at 530 nm and the long (L) at around 560 nm.

Studies show that the M and L pigment genes sit next to each other on the X chromosome while the S pigment gene lies on chromosome 7. The M and L pigments are almost identical and the observed difference in spectral sensitivity deriving from substitution in just three of the 364 amino-acids. However, the S pigment gene's amino-acid sequence seems to be only distantly related to the M and L ones.

Almost all vertebrates have genes with sequences that are very similar to that of the human S pigment, implying that some version of a shorter-wavelength pigment is an ancient element of color vision. Interestingly, even though relatives of the two longer wavelength pigments are widespread among vertebrates, the presence of both M and L-like pigments has been seen only in a sub-set of *primate* species - a sign that this feature may have evolved more recently.

In fact, the disparity in color vision observed between the so-called New and Old World primates, African versus South American species, provides a unique window into the evolution of color vision in both groups.

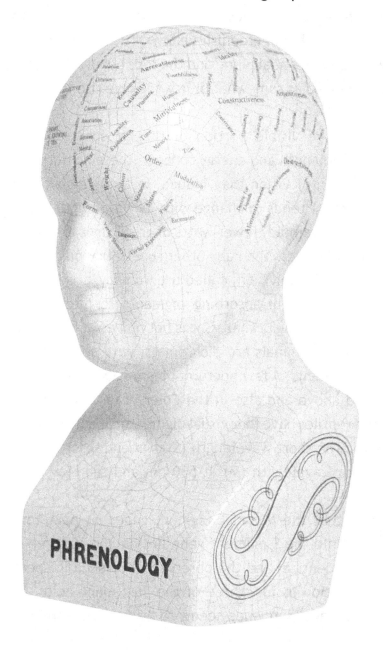

# Temporal events in Vision

The vertebrate visual system has evolved over *billennia* (1000 millennia) to allow animals to gather information about the external physical world with which it has only indirect contact. It utilizes photons, a key component of the sun's radiant energy to provide it with this information. Visible light energy covers a broad dynamic spectrum from approximately 390 to 760 nm, a range of luminance that the visual system of most primates has adapted to in order to take advantage of its availability here on Earth.

When light encounters regions of varying density it slows down or refracts depending on the optical density of the intervening medium. Always in a hurry to get through obstacles, the light beam chooses a path that requires the least transit time. This time interval determines the refractive index of the medium. By definition, light travels the fastest in air and holds the value of one. Consequently, a denser medium like water or glass will have a refractive index of higher than one i.e. more transit time.

According to electromagnetic theory, refraction may be represented by light waves of a certain frequency entering a medium and setting up oscillations within it.

(**Author's Note**: When light passes into a denser medium, its velocity changes but frequency remains constant i.e. its wavelength changes proportionately. Since it is easier to measure wavelengths than transit time, the oscillations are defined by the intensity of the source.)

Max Planck postulated that the light sources or oscillators do not give out light in a continuous stream but in discrete packets of energy called quanta. Increasing light intensity increases the number of quanta but their energy is fixed by frequency since all travel at the same velocity. As a simple visual, the quantum resembles a "missile" particle traversing a trajectory (wave) through the optical medium. So, depending on the computation criteria, interference or interaction of these quanta with the media molecules can be depicted as a graph or dot matrix pattern.

The very act of seeing involves both aspects: waves pass through ocular media whose routes are altered by eyewear interference. The particulate photons upon reaching the retina alter the photo-pigments in the receptor cells to induce nerve impulses in retinal ganglion cells. These encoded signals then travel along the optic nerve and surrounding neural tissue into the cortex for further evaluation, which results in the formulation of a cohesive percept of the visual field.

Another class of specialized visual properties deals with both rapid and slow changes of radiant energy as a function of time. The visual system responds to such changes in a manner that allows nearly instantaneous interpretation of a rapidly changing environment or scenario. For example, a gazelle being ambushed by a leopard either needs to get out of the way or become dinner.

Through an in-built editorial capacity, it avoids being overwhelmed by an excessive amount of information. Subjectively, visual imagery appears to be stable and the objects tend to move smoothly in time and space. Yet most of what we see is only selected portions of a potentially infinite variety of images taken from our surroundings.

The visual system periodically samples the images cast on the retina. It then stores, integrates, differentiates, erases and performs other operations resulting in the perception of apparently stable scenes.

Since the light energy arriving at the eye varies continuously in time and place, interpretations of these variations must take place in a nearly synchronous fashion. The temporal responsiveness of the visual system is necessarily limited i.e. a finite amount of time is required to collect and process the information in any image.

Although visual experiences generally tend to conform well to the spatiotemporal order of the external physical world, they are never a perfect representation of reality especially if your optometrist was sleeping and your refractive error was not adequately corrected. This is a result of the limited responsiveness of physiological mechanisms in the visual system.

These mechanisms edit all visual information, condensing or discarding redundant or irrelevant features while enhancing and retaining relevant ones. Functionally useful visual information consists of variations of light in either time or space. That which is visually significant is the presence of "something different", a change in the image. Contrasting boundaries of light in the retinal image and changes in their location or magnitude are all that is relevant.

The visual system responds appropriately by acting as a differentiator and as an integrator, separating the chaff from the dross. It fills in apparent voids in visual content and serves to interpret a constantly changing pattern of stimulation, using a time-based, continuous search for invariances and orderly relations within retinal images.

The basic law underlying temporal integration is similar to the Bunsen-Roscoe law of photochemistry, which explains that since some 500-odd rod cells connect to each optic nerve fiber with a receptor field of about 10 arc-seconds, the cascade of photons should arrive at these cells within a 100 msec period of time so as to not dissipate their effect.

Since photochemical changes initiating vision are almost instantaneous, the longer time-sensitivity interrelation for the process of sight is known as Bloch's law. It implies that a visual threshold may be reached by a particular number of quanta (photons), irrespective of their time distribution. It invokes a more graded action potential rather than the usual all-or-none responses commonly observed throughout the CNS.

A minimum number of quanta on the intensity side and a minimum duration on the temporal side impose the limits (critical duration).

Critical duration implies that the eye can summate quantal effects. For stimuli lasting longer than the critical duration, intensity alone is important. Temporal summation depends on both photochemical and neural factors. The eye functions over a large range of luminance levels from one to $10^{10}$. The cones mediate photopic vision, the rods, scotopic and both function best under intermediate mesopic ranges of illumination.

Interestingly, this delineation of neural 'territory' is best depicted by the "discontinuity" in the graph when various visual functions such as acuity, adaptation etc are plotted against luminance intensity. The transition from one type of vision into the other usually occurs at around 10 candelas/$m^2$ of luminance.

The most effective wavelength is 510 nm (max rhodopsin sensitivity) and triggered by only 2 to 5 photons/msec arriving at the rod cell. Analogous to the ripples produced on a quivering pond surface by a thrown pebble, one photon-pebble is sufficient to initiate the visual rhodopsin cascade, wherein two to five surrounding rods, are also stimulated to signal its arrival. The cortex then interprets their *collective* input as a visual sensation.

The visual "brain" is very fast and tachistoscopic experiments on macaques as well as human subjects reveal that we can distinguish animal pictures from non-animal ones within 150 msecs. The critical duration tends to vary with background luminance and the adaptation state of the eye: as little as a 100 msec for foveal tasks to 400 msec for more complex ones such as interpreting form, motion or object recognition.

Brief stimuli are not perceived as evolving in time. For example, when two short events (a red light flash followed by a green, 10 msecs apart) follow each other, a single "yellow" flash is reported rather than a green or red. This indicates that there is not a full "masking" of the first stimulus (red) by the later flash (green) but that the visual brain seems to be melding them into a single "hybrid" percept.

Behaviorally, the interval during which such 'backward' masking is effective can be extended up to 100 msecs but with some residual neural processing having already taken place i.e. reporting a 'yellow' light instead of a red or green one, in the example above.

This finding implies that awareness does not commit itself to a percept until about 80 msec after the stimulus event has taken place.

Intermittent light from strobes or fluorescent bulbs produce a flickering effect until at a certain frequency the sensation becomes continuous. This readily measurable threshold called the critical fusion frequency (CFF) varies directly with the log of luminance (Ferry-Porter Law). Its brightness at fusion matches a steady light with the same average luminance (Talbot-Plateau Law).

Perception of movement is essential for survival in the animal kingdom. The lowest velocity, at which motion can be perceived, ranges from one to eight arc-seconds, depending on the background against which it is being detected. In contrast to such real motion, is the perception of apparent motion due to sequential retinal stimulation (phi phenomenon)?

Motion pictures, television and various neon signs fit this category.

It is not clear if similar neural mechanisms underlie both types of perception.

# The Visual System: An Overview

# Retinotopic Maps & Visual Fields

Vision in most diurnal creatures like the primates and also humans is considered to be the predominant sense. While other senses such as hearing and touch are essential for our well being, vision plays a key role in how we extract meaningful information from the environment and formulate our perception of "reality".

The visual field is the part of the surroundings from which the eyes can perceive light. Together, the two eyes cover a large area, almost 270 degrees all around with each eye responsible for almost 140 degrees by itself.

Since each eye is approximately an inch in diameter, its nasal curvature looks out onto the temporal field out there and its temporal curvature views the nasal field. Similarly, the superior or top half of the retina looks at the ground and the inferior half views the sky.

Out of the 270 degree field that was mentioned earlier, the lateral 30 degrees of the visual field on either side is seen by one eye only (monocular zone). The so-called binocular zone, straight ahead, comprises the rest and any "damage" affecting the retina within this zone is usually not noticed by the patient. For instance, the blind spot in each eye.

Since each point in the retina corresponds to a specific location in the visual field, it becomes easy to pinpoint observed visual field defects when any regions of the retina are damaged or vice versa. In addition to the extremely precise retinotopic arrangement of the visual pathways, another important feature is that the density of retinal ganglion cells is considerably higher in the central region (macula) than in the periphery.

This implies that the macular area and consequently the central visual field is over-represented in both the LGN and visual cortical maps. In a way this makes a lot of sense as in predatory animals, it is very important to gather as much usable information of prey items in the central visual field. Precise control of eye movements also ensure that light from the central field always falls within the macular region for rapid and efficient analysis of visual information in "real" time.

# Binocular Vision

Just because we have two eyes does not mean that we possess binocularity. There are quite a few mammals and birds with overlapping visual fields that are not truly fulltime binocular like the primates. Each eyeball needs to be wired into the visual cortex in a special way for the brain to fuse the retinal 'images' into a stereoscopic whole.

We know this because about two percent of the population can see well with each eye individually but the random-dot stereograms (a test of binocularity or stereopsis) remain flat. Another small percent possess poor degrees of stereopsis ranging from intermittent to flat fusion. This variety of 'stereo-blindness' appear to be genetically determined, suggesting that the visual cortex is 'assembled' piecemeal and not as a complete module.

Stereovision is not present at birth and can be permanently damaged in children or young animals if even one of the eyes is temporarily deprived of sight for a few days. When infants are born every neuron in the receiving layer of the visual cortex "adds up" the inputs from corresponding locations in the two eyes rather than keep them separate. The brain is unable to form a right-left spatial map of each eye individually.

Around the two to three month stage, the visual cortex becomes sophisticated enough to differentiate from which eye the input is originating. As soon as this happens, a sort of "rivalry" occurs between the two eyes as their spatial maps now have a built-in bias. Each eye is interpreting a different aspect of the visual field with a large central area of 'overlap'. This mismatch called binocular rivalry forms the basis of random-dot acuity in primates.

Once the brain has segregated the left eye's image from the right eye's, subsequent layers of neurons can compare them for the minute disparities that signal depth. When the random dot pattern, usually generated by a computer, is seen by the viewer, the tiny disparities built into the pattern are interpreted by the brain as a picture containing various degrees of depth i.e. three dimensions rather than just two.

Suddenly, as the numerous inputs to the visual cortex coalesce the flat 2D dotted image begins to morph magically into a robust 3D one, sometimes so rapidly that the viewer may take a step back or bob his head around in curious surprise.

Just as Hollywood makes use of the persistence of vision to convert flashing still pictures into movies, this peculiarity of the visual system is exploited in the various Magic eye pictures and so called "optical illusions" like the Necker Cube, mother-daughter, vase-human face profiles etc. popularized in sundry magazines/books.

Stereovision is information processing that we experience as a particular flavor of consciousness, a subtle connection between mental computation and awareness. Even

mediated by the extra-ocular and ciliary muscles of the two eyes, it is sensitive to

experience. That is why, if you have seen a certain stereogram before, the image pops out instantly and gets more detailed as you scrutinize it.

The discoveries regarding this 'tunability' of stereopsis in different species affords a novel way of thinking about learning in general. According to Steven Pinker of MIT:

"Learning is often described as an indispensable shaper of amorphous brain tissue. Instead it might be an innate adaptation to the project-scheduling demands of a self-assembling animal". In other words, Pinker feels that the genome builds as much of the organism as possible but leaves the final details of physical development to an information-gathering mechanism.

Several brain researchers have found that this fine-tuning of the neocortex occurs at the local level (lateral geniculate body/ superior colliculus) where the receptor organ (eye) is 'hardwired' to the brain as and when needed during development. This process may have evolved in response to identifiable selection pressures in the ecology of our ancestors. It provides reasonable solutions to unsolvable problems by making tacit assumptions about the world out there.

Most neuroscientists recognize that at the highest levels of cognition, we consciously plod through various steps and invoke rules we learned in school or discovered ourselves. The 'mind' is compared to a complex production system with symbolic inscriptions embedded in a huge memory bank with a bevy of minions that carry out the procedures. At a lower level, the inscriptions and rules are implemented in minor neural networks, which respond to familiar patterns and associate them with other patterns.

However, the boundary is in dispute. Do simple neural networks handle the bulk of everyday thought, leaving only the products of book learning to be handled by explicit rules and propositions?

On the other hand, are the networks more like building blocks that are not humanly smart until assembled into structured representations and programs?

It is often heard that animals are not well engineered at all. The process of natural selection is hobbled by shortsightedness, the dead hand of the past and crippling constraints on what kinds of structures are biologically and physically possible. Much unlike a human engineer, natural selection is incapable of good design. Animals are clunking jalopies saddled with ancestral junk and more often than not blunder into barely serviceable solutions.

People are so eager to believe this claim that they seldom question the whereabouts of this hypothetical engineer that is not constrained by availability of parts, manufacturing practicality or the inherent laws of physics. Sure natural selection does not have the foresight of engineers but it also dispenses with their impoverished imagination, mental hang-ups or rabid conformity to bourgeois sensibilities and stakes in profitability.

Guided only by whatever works, the relentless process of natural selection can invent brilliant creative solutions. For millennia, biologists have delighted in the ingenious contrivances found in the animal world: the biomechanical perfection of cheetahs, the infrared sensors of snakes, the ultrasound echolocation apparatus of bats and dolphins, the superglue of marine barnacles, and the super strong silk of spiders, the prehensile tails of macaques or the manual dexterity of humans.

After all, entropy and the more malevolent forces of nature are constantly gnawing at an organism's right to survive, culling the herd with relentless impunity and avidly decimating the outcomes of sloppy engineering. Gould has emphasized that natural selection has only limited freedom to alter basic body plans. Much of the plumbing, wiring and morphology of the primates, for example, have been unchanged for millions of years.

Presumably, they have emerged from embryological 'blueprints' that cannot easily be tinkered with. However, the primate body plan can accommodate marmosets, monkeys, tarsiers, lemurs, chimps, orangs, gorillas and hominids. The similarities are important but so are the differences. Developmental constraints are only able to rule out broad classes of options. They cannot induce a fully functional organ to come into being.

Natural selection can only pick-n-choose from alternatives that are growable as carbon-based living stuff. Another widespread misconception is that if an organ changed its function in the course of evolution, it did not evolve by natural selection. However, a detailed look at the animal world reveals that many organs that we see today have maintained their original function.

For example, the eye was always a visual organ whether it began as a light-sensitive spot or ended up as an eagle eye. Others did change their function. Pectoral fins of fishes became the forelimbs of horses, the flippers of whales, the wings of birds, digging implements of burrowing animals or the brachiating arms of orangutans. More often, before an organ was selected to assume its current job, it was adapted for something else.

The delicate chain of bones forming the middle ear in primates actually began as jawbones in reptiles. Reptiles often sense vibrations by lowering their jaws to the ground. As the mammals came into their own after the devastation of reptiles in the cosmic holocaust of 65 million years ago and began to occupy diverse ecological niches, this function gradually transformed the lower jaw into a much-reduced vibration-sensing organ in mammals. Over millennia, these tiny bones became nestled in the temporal bone forming the middle ear of primates.

There is nothing mysterious about the evolution of bird wings either. Half a wing will not enable you to soar like an eagle but it will let you glide gently from tree to tree or parachute you down from a tall tree to the forest floor, to help cushion your fall. Flapping their vestigial wings do afford ostriches and the lowly chicken to get the hell out of the predator's way in a hurry apart from raising a dust cloud to also foil the cat's evil intentions.

Electrophysiological recordings in the macaque, clinical reports and brain imaging in humans implicate the posterior parietal (PP) cortex in combining and expressing position information and relating it to action. The PP is subdivided into numerous functionally distinct regions where the neuronal responses that are neither purely sensory nor purely motor but a blend of both.

Single-cell experiments indicate the PP is involved in such diverse functions as analyzing spatial relations among objects, controlling eye-hand coordination and determining where to allocate visual attention next. PP is an important conduit for action-related information. The output pathways include direct projections from layer 5 of the PP to the spinal cord, brainstem and reciprocal connections to the premotor and prefrontal areas of the cortex.

# Fixational Eye Movements

As you weave your way through crowds of commuters in Grand Central Station, your torso, legs and arms continuously adjust themselves so that you remain upright and avoid bumping into anybody. Even though you are quite oblivious of such action corrections, numerous neural networks mediate balance and body posture in real time via continuously updated information from many modalities, not just vision.

By just looking carefully at the various animals we can see that eye positions change from a lateral aspect where each eye is visualizing its own panorama, to eyes with higher degrees of overlapping fields. When the eyes are in the primary position of gaze i.e. viewing objects at optical infinity, the binocular field can be considered as simply the amount of overlap of each monocular field.

For front-facing eyes, the visual axes are almost parallel and the monocular field overlap quite maximum. Some animals possess rather prominent snouts, bills or other facial features that may limit the binocular field. Only simians and humans appear to have developed an ability to converge the eyes so as to bifixate objects as close as a few inches from the bridge of the nose.

The eyes and their distinct patterns of movement are a fascinating source of information. A total of six eye muscles are responsible for rotating the eyeball in several predetermined or learned patterns by the brain. A saccade is a rapid (usually measured in milliseconds) movement of both eyes yoked together to "zoom" in on an object of interest. Pursuits, on the other hand, are smooth continuous tracking movements of both eyes used to keep the object of interest in constant view while both the viewer and object are in motion.

When the eye movement is off target (overshoot or undershoot), a corrective saccade of small amplitude is triggered to bring the target back onto the fovea. Contrary to popular opinion, the eyes move all the time even when they are "locked-on" to a target. These tiny "jiggles", known as micro-saccades take place within Panum's space (explained in another section). These micro-saccades prevent the retinal image from fading or going away.

The intervals between saccades are brief, as short as 120-130 msec. This corresponds to the minimum time needed to process visual information during fixation. Saccades are usually mediated by the superior colliculi of the brain while the smooth pursuits are coordinated by parietal and/or prefrontal cortex.

If eye movements are prevented or the extra-ocular muscles are tranquilized during surgery, vision rapidly fades into an indistinct "fog". It is often assumed that such "fading to black" is a purely retinal phenomenon caused by the stability of an image on the photo-receptors.

Unfortunately, very little is known about the neuronal basis of fading but it may have something to do with the refractory periods of the ganglion cells and the constant need to keep neural impulses flowing through the optic nerve into the LGN to keep the "movie" alive.

The stability and sharpness of the visual world during eye movements is a direct consequence of numerous processes including saccadic suppression, a mechanism that interferes with vision during eye movements. This can be experienced by looking in a mirror and fixating each eye alternately. You will never catch your eyes in transition when going back and forth with each eye.

It seems that during the transition phase, vision is partially shut down by the brain. This eliminates blur and the queasy feeling of a world in motion around you. Why then isn't everyday continuous vision "gappy" and broken up by sudden saccades or blinks? Some researchers have postulated the presence of a trans-saccadic integration mechanism that fills in these intervals with a "fictive" movie bridging the retinal images just before and after the blink or saccade.

The extraocular muscles (EOMs) responsible for moving the eyes receive their innervation from the nuclei of the III, IV and VI cranial nerves. The control system coordinates the activity of the alpha motoneurons in these nuclei and premotor networks interconnecting areas in the cerebral cortex and brainstem carry out the task. To enable such coordination, the nuclei are also interconnected by numerous fibers forming the medial longitudinal fasciculus (MLF).

The control system uses sensory information from several sources. For example, the retina informs about whether the image is stationary or slipping in real time, the vestibular nuclei monitoring the degree of head tilt or rotation along the Y & Z axes and the impulses coming from the proprioceptors in the eye muscles relaying information about the actual eyeball coordinates within the orbits. All this sensory information is integrated by the brain and transformed into a single volley of motor signals that help fine-tune EOM activity.

There is an important difference between the control of EOMs and other muscles subjected to precise voluntary control e.g., finger movements of a pianist:

the nuclei of the EMs, in contrast to the spinal motoneurons, receive no direct fibers from the cerebral cortex.

The central control is exerted via premotor networks in the brainstem. In fact, eye movements are voluntary only in a limited sense. We cannot move the two eyes independently of each other like the chameleon and the slow pursuit movements of which we are aware can only be performed when tracking a moving target.

Different central nervous networks control each of the numerous movements that the eyeballs perform every day although all converge on the alpha motoneurons controlling the EOMs. Recent studies show that several regions, for example in the cerebral cortex and the cerebellum participate in the control of both saccadic and pursuit movements.

Signals informing about desired eye position, actual position, retinal slip and position of the head are integrated in the reticular formation (RF) lying adjacent to the eye muscle nuclei in the brainstem. The lateral regions of the pontine nuclei mediate voluntary pursuit movements and the slow-phase components of OKN (optokinetic nystagmus).

Vertical eye movement control mechanisms are found in the mesencephalic RF close to the oculomotor nucleus. The premotor "gaze centers" are supplied by the vestibular nuclei, the superior colliculus, the cerebral cortex and others with relevant information named above.

At least two regions of the cerebral cortex are closely involved in the control of EMs: the frontal eye field is primarily related to initiation of saccades whereas several smaller areas in the parieto-temporal region mediate pursuits.

Your eyes never stop moving, even when they are apparently "fixating" on a target. They still jump and jiggle imperceptibly in ways that turn out to be essential for vision. These 'micro-saccades' help to refresh the retinal image of the object of regard, with each excursion, onto each eye's fovea.

In the last few years, vision scientists have detected tell-tale patterns of neural activity that correlate these micro-saccades with human perception and awareness. What is more, instead of being random movements, they may point to where your "mind" is secretly focusing - even if your actual gaze is directed elsewhere - revealing hidden thoughts and intentions.

Indeed, animal nervous systems have evolved to detect changes in the environment, as quickly spotting differences ensures survival. Any motion or change in the elements of the visual field signal approaching food or predator.

Unchanging objects generally pose a threat, so animal brains and visual systems did not evolve to notice them. For example, a frog will tend to ignore edible but dead flies suspended motionless all around it in a bizarre experiment, thinking they do not exist (neural adaptation), but will immediately flick its tongue out towards one that is moved up and down by the technician.

In humans, the small fixational eye movements help shift the visual scene across the retina, prodding visual neurons into action and thus counteracting neural adaptation. Thus, no matter how hard you may avert your eyes from that last piece of cake or an attractive man or woman near you, the rate and direction of your micro-saccade betray your true attentional focus.

# Fusion, Stereopsis & Panum's Space

Normal binocular vision implies binocular single vision (fusion) and a high level of stereo-acuity. Stereopsis is the binocular perception of depth enabled by the two eyes viewing the external world from disparate vantage points. The depth estimates in stereoscopic vision are not absolute or egocentric estimates but depend on the fixation point being either nearer or farther away.

Stereoscopic depth perception requires the correlation of visual information from both eyes that are frontally placed to allow a considerable overlap of the binocular visual fields. It also requires yoked eye movements so that the objects stimulate corresponding retinal points and a semi-decussation of the optic tract to permit the interaction of inputs from corresponding regions of each retina. Retinal points in the two eyes are said to be corresponding if when stimulated separately they appear to have the same visual direction.

The term horopter, (Greek horos, boundary), defined as the locus of points seen as single with the two eyes was originally introduced by Aguilonius in 1613 to explain the concept of corresponding points in the two retinas having the same monocular visual direction. It was conceived as a surface in visual space. Since that time, the application of the horopter concept has shifted away from a special concern for binocular single vision toward more general problems of stereopsis.

Both, from a theoretical and a physiological point of view, the horopter is best defined as the locus of maximal stereoacuity. In line with this definition, the horopter can also be conceptualized as a zero-disparity reference plane that contains the fixation point and relative to which stereoscopic depth estimates are made.

Wheatstone, by his invention of the stereoscope in 1838, was the first to recognize that disparate retinal images provide the essential cue for binocular depth perception. His idea was basically simple. Take two pictures from slightly different vantage points similar to your two eyes, place the right picture in front of the viewer's right eye and do the same with the left. The brain assuming that the two images are from the real world ends up combining them into a cyclopean image and the binocular parallax provides the three-dimensionality or depth to the percept.

However, as is evident from our earlier discussion on binocular vision, all humans fail to see these stereograms in stereo. Stereovision gives information only about relative depth i.e. how much the object of regard is in front or behind the fixation point (plane). This is mediated in the brain by integrating information from the focusing reflex mechanism (ciliary muscle and crystalline lens) and vergence movements (extra-ocular muscles), both of which are yoked.

In order to see the stereogram in full 3D glory, these two mechanisms need to be uncoupled. Viewers resort to various ways to achieve this uncoupling: some cross their eyes at an imaginary distance in front of the stereogram and "free-fuse"; some intentionally view the full picture as if through a mild "fog" by going into a "soft-focus" mode etc.

Another thing Wheatstone recognized was that stimulation of disparate or non-corresponding retinal points could still produce single vision, effectively contradicting the theory of corresponding retinal points.

Panum, in 1858, proposed that the singleness of binocular vision was not confined to a single surface but extended over a volume in space. He argued that for any point on one retina, there is a small circle or area of points on the other retina, stimulation of which will lead to fusion of two monocular inputs.

Thus, binocular single vision is not limited to the immediate vicinity of the horopter, but extends for a small distance proximal and distal to it. This space came to be known as Panum's fusional area beyond which the subject will see two images (physiological diplopia) of the object rather than just one.

Studies show that at the fixation point the extent of Panum's area is approximately 15 minutes of arc, well within the normal range of the human foveola. These results are also similar to those inferred from studies of fixation disparity.

Fixation disparity is a variant of normal binocular vision. The disparities are usually found to be small, ranging from -5 to +3 minutes of arc. In any case, they cannot be larger than Panum's fusional area if diplopia is not to result. Furthermore, they found that the extent of the fusional area was the same for vertical or horizontal disparities.

The extent of Panum's fusional area increases with increasing eccentricity away from the fixation point. Interestingly, Panum's area depends greatly on the class of stimulus and upon achieving binocularly; stabilized global stereopsis may stretch to almost $2^0$ in the horizontal direction without loss of stereopsis or fusion. It seems therefore that there is a fairly large tolerance for disparate images once fusion has been obtained.

# Binocular Rivalry

Consider the twelve lines making up the Necker cube. Due to the inherent ambiguity of inferring its 3D shape from a 2D drawing on a plane sheet of paper, the lines of the cube can be interpreted in two ways, each differing in their orientation in space. Without perspective and shading cues, either percept is possible. Even though the line drawing does not change, conscious perception does flip back and forth between the two interpretations in what is known as a bistable percept.

Interestingly, you never see the cube suspended in space, halfway in either of the two positions nor do you see a blend of the two figures. Your mind cannot simultaneously visualize both shapes. Instead, each configuration vies for perceptual dominance. This is but one manifestation of a general phenomenon that in the presence of ambiguity prefers to accept a single interpretation, which may change with time. This aspect of experience is sometimes referred to as the unity of consciousness.

In the course of everyday life, your eyes are constantly presented with similar but non-identical views of the world. The brain can extract sufficient cues from the small discrepancies between these two images to discern depth. Lets take another example from Optometry called the Worth's Four Dot. Here, the subject dons a pair of red-green glasses and views a flashlight with two red dots at the 3 and 9 o'clock positions and a green dot at the 12 o'clock position. The 6 o'clock position has a clear dot.

When the subject is asked to report the color of the 6 o'clock clear dot, they invariably report a red or green dot depending on which of their eyes is dominant or an alternating red-green configuration or a yellowish dot if unable to decide. The two red-green percepts tend to alternate in this manner indefinitely but only one is seen while the other is suppressed.

During this so-called binocular rivalry, the two enter and depart from consciousness in a never-ending dance. What you see is not a superposition or blending of the two images but just one. It may belong to the dominant eye, which the brain uses preferentially for most viewing tasks. The duration of these dominance periods i.e. how long either one is visible, varies considerably across subjects and trials.

Binocular rivalry can be thought of as a reflexive alternation between percepts that can be influenced though not completely abolished by sensory or cognitive factors. At the neuronal level, rivalry was long believed to be due to reciprocal inhibition between populations of cells representing input from the left and from the right eyes. One coalition fires away, preventing the other from responding. As this inhibition fatigues, the other group eventually dominates.

Recent psychological and imaging evidence has suggested that this automatic switching is complemented by active processes linked to attention. Mechanisms located in prefrontal and parietal areas can bias the system toward one or the other coalition. This enables the chosen coalition to build up sufficient strength to dominate and to widely distribute its informational content, bringing that image to consciousness.

Where in the brain does the fight for dominance occur?

The retinal neurons are not influenced by the percept. They are driven exclusively by the photoreceptor input. Perceptual modulation could occur as early as the lateral geniculate nucleus, halfway between the retina and primary visual cortex. However, recordings from geniculate neurons have shown that their firing rate is indifferent to whether a monkey saw a rivalrous or a non-rivalrous stimulus. The interplay between dominant and suppressed stimuli, therefore, occurs in the cortex.

The cortical sites underlying binocular rivalry in monkeys were explored by numerous researchers, at MIT and elsewhere. They found that the majority of cells in the primary (V1) and secondary (V2, V4 etc.) visual cortices fired with little regard for the ebb and flow of perception. Largely, a neuron increased its activity to the stimulus in one eye no matter what the macaque monkey saw.

Only six of the 33 cells were somewhat modulated by perception. The majority of V1 cells fire no matter whether the monkey sees one or the other stimuli. This implies that exuberant cortical activity does not guarantee a conscious percept i.e. not just any cortical activity contributes to consciousness.

What about intermediate cortical areas? Do they mediate binocular rivalry?

The neuronal response patterns in areas V4 and MT are more varied than those in V1. About 40% of V4 cells are correlated with the animal's behavior, that is, with its (assumed) perception. The firing profile of many of these cells indicates that they change their output primarily during transitions when the percept changes over from one image to another.

One plausible conclusion is that the coalitions in these intermediate cortical areas are competing against each other attempting to resolve the ambiguity imposed by the two disparate images. At some point, the winner is established and its identity (and probably that of the loser) is signaled to the next stages of visual analysis.

When the researchers recorded from cells in the inferior temporal (IT) cortex and in the lower bank of the superior temporal sulcus (STS), that delimits IT on its upper side; they found that the competition between the rivalrous stimuli was resolved.

Nine out of ten cells fired in congruence with the macaque's percept i.e. whenever the monkey saw the preferred image, the neuron fired. When the other image dominated, the cell was mute. Unlike the situation in V4 and MT, no IT cells signaled the suppressed and invisible image.

The IT cortex and neighboring regions not only project to the prefrontal cortex but also receive input from it. What is the role of this feedback in binocular rivalry and related phenomena?

To explore these questions, Nikos Logothetis of the Max Planck Institute turned to the analysis of binocular rivalry in macaques. He trained his monkeys to press one of two levers to indicate which object was being perceived. To ensure that the monkeys were not responding randomly, he included non-rivalrous trials in which only one of the objects was presented.

Recordings were then made from single cells in various areas of the visual cortex. Within each area, he would select two objects, only one of which was effective in driving the cell. In this way, he could correlate the activity of the cell with the animal's perceptual experience.

As his recordings moved up the ventral pathway, Logothetis found an increase in the percentage of active cells, with activity mirroring the animal's perception rather than the stimulus conditions. In V1, the responses of less than 20% of the cells fluctuated as a function of whether the monkey perceived the effective or ineffective stimulus. In V4, this percentage increased to over 33%.

In contrast, the activity of all cells in the visual areas of the temporal lobe was tightly correlated with the animal's perception. Here the cells would respond only when the effective stimulus was perceived. When the monkey pressed the lever indicating that it perceived the ineffective stimulus under rivalrous conditions, the cells were essentially silent.

In both V4 and the temporal lobe, the change in cell activity occurred in advance of the animal's response that the percept had changed. Thus, even when the stimulus did not change, an increase in activity was observed prior to the transition from a perception of the ineffective stimulus to that of the effective one.

These results suggest a competition during the early stages of cortical processing between the two possible percepts. The activity of the cells in V1 and in V4 can be thought of as perceptual hypotheses, with the patterns across an ensemble of cells reflecting the strength of the different hypotheses.

Interactions between these cells ensure that by the time the information reaches the IT lobe, one of these hypotheses has coalesced into a stable percept. Reflecting the properties of the real world, the brain is not fooled into believing two objects exist at the same place at the same time.

# Optical Illusions

How can the visual system compute the most probable 3D state of the world from the 2D input it receives from each retina?

First, it searches through its memory banks for an object that has the greatest likelihood of producing those lines or sketch based on the Occam's razor principle of parsimony, and

Second, it computes the probability of that particular object fitting the context of that visual scene (the proverbial zebra in the backyard scenario)

For example, a set of parallel lines are seldom an accident, for non-parallel lines in the world rarely project almost parallel lines in an image. Many laws in the natural world give it nice, analyzable shapes. Motion, tension and gravity make straight lines. Gravity makes right angles. Cohesion makes smooth contours.

However, vision is not a Cartesian theater designed by our awareness. We experience only the scene unfolding directly in front our eyes, the rest of it appears as a vague, peripheral smear of an ethereal existence. Our eyes flit from spot to spot several times a second forming an etch-n-sketch like image of the surroundings. We see in perspective. We are aware from experience that moving objects loom, shrink in size and foreshorten.

Ambient light also influences a lot of the visual field and certain built-in laws of constancy guide the visual cortex in arriving at appropriate rationales for what it is seeing. Since we are used to seeing surfaces rather than volumes, we have an almost palpable sense of surfaces and the boundaries between them. Some of the most famous illusions in optometry stem from the brain's unflagging struggle to carve the visual scene into surfaces and to decide where they would be likely to fit in the real world scenario.

For example, take the Kanisza triangle or the Rubin face profiles/vase illusions. The reason the viewer has such a hard time deciding between what is seen and what is actually there is because the brain knows intuitively that the picture is flat or 2D but is simulating 3D objects.

So what does the brain end up doing?

Marr, with tongue-in-cheek I am sure, chose to call it a $2^{1/2}D$ sketch while other researchers call it a visible-surface representation.

The information in this modified 2D array is specified in a retinal frame of reference, if you will, which simulates a coordinate system centered on the viewer. There are a series of reciprocal reference frames like the vestibular system, cerebellum and basal ganglia or the superior colliculus that compensate for movements of the head and body. They provide each bit of surface in the visual field a fixed address relative to the field, which stays the same as the body or eyes move.

Irv Biederman, a psychologist, has fleshed out Marr's ideas by invoking an inventory (24) of simple geometric forms called geons, which tend to be combinatorial. His theory suggests that at the highest levels of perception, the mind sees objects and parts as idealized geometric solids. These geons are not good for everything for many natural objects such as mountains and trees have complex fractal shapes.

In a highly social species like ours, or even the great apes taken together, face recognition and being able to "read" facial expressions is very important. Some studies suggest that face recognition may even use distinct parts of the brain. One famous set of experiments pointed to mental rotation of the object in view as central to successfully solving the dilemma such face or object recognition tasks pose to our brain.

Cooper and Shepard demonstrated that the brain rotates objects in mental space in order to decide its handedness, right or left or to figure out if it is a mirror image of the viewed object. Mental imagery drives our thinking about objects in space but we do not use such imagery to rearrange furniture or figure out where a person is sitting in a huge mansion. Many creative people claim to see the solution to a problem by visualizing it in an image.

For example, Crick and Watson mentally rotated models of what was to become the double helix; Einstein imagined what it would be like to ride on a beam of light. Painters and sculptors try out various ideas in their minds and novelists devise scenes and criminal plots in their mind's eye before putting it down on paper. Images drive the emotions as well as the intellect.

However, what is a mental image? Brain scientists call it a topographically organized cortical map i.e. a patch of cortex in which each neuron element responds to contours in a specific portion of the visual field. The primate brain has at least fifteen of these maps and truly, they are pictures in the head. As explained in previous sections, ambient space out there is represented by a topographical or contour map of the visual terrain.

The brain is also geared to interpret or integrate information flowing down from the limbic system with its memory stores into these mental images. Imagery can affect perception in gross ways too. Mental images of lines make it easier to judge alignment and can even induce visual illusions.

# The Primary Visual Cortex or V1

The cerebral cortex can be subdivided into the phylogenetically older olfactory (smell) and hippocampus (memory) cortex and the newer neocortex. This multilayered "cake-like" structure crowning the rest of the brain is only found in mammals. The human neocortex and its diverse connections occupy about 80% of the total brain volume. Quite different from other brain structures such as the thalami, basal ganglia or brainstem, the neocortex is a vast sheet of neural tissue.

It is highly convoluted and possesses a laminated substructure. The size of one cortical sheet varies across species, ranging from around 1 cm$^2$ in the tree shrew to about 100 cm$^2$ in the macaque monkey, 1000 cm$^2$ in humans and dolphins to several times larger in certain whales. Think of its crinkly appearance as 2 thick pancakes, 35 cm in diameter, crumpled up and stuffed into your skull. The exact pattern of cortical indentations (gyri and sulci) is as unique as an individual fingerprint.

There are many types of cortical neurons. Based on the laminar position of the cell body, dendritic morphology and axonal target zones, about a 100 cell types can be distinguished. Pyramidal cells tend to predominate with extensive inter-connections both inside and outside the cortex.

While the receptive field structure of retinal and geniculate cells is relatively stereotypical, the cortical cells display an amazing variety of selective responses to motion, color, orientation, depth and other stimulus features. Their non-classical receptive field extends far beyond the confines of the region in space that directly excites the cell. It provides the context within which any single visual stimulus is placed.

In humans, much of V1 is buried within the calcarine fissure on the medial wall of the brain as corresponds to Brodmann's area 17. The location and orientation of this fissure can vary between individual brains and the outside world is mapped onto V1 in a topographic manner with neighboring locations in the visual field projecting onto nearby locations. Optometrists refer to this spatial organization as a retinotopic map.

The visual cortex contains multiple, superimposed charts or maps for the position, orientation and direction of motion of stimuli, ocular dominance and color. Are these maps related in some way to each other or are they random? It is unclear at this point. What is clear is that visual processing along these two new pathways is designed to extract fundamentally different types of information.

The ventral or occipito-temporal pathway is specialized for object perception and recognition for determining what it is we are looking at. The dorsal or occipito-parietal pathway, on the other hand, is specialized for spatial configuration between different objects in a scene.

"What?" and "Where?" are the two basic questions to be answered in visual perception. Not only must we recognize what we are looking at but also we need to know where it is in order to respond appropriately.

Upon emerging from V1 – and now known as the vision-for-perception (ventral) and the vision-for-action (dorsal) streams – they flow toward the prefrontal cortex.

The ventral stream passes through V2 and V4 into IT and projects from there into the ventrolateral prefrontal cortex. This pathway is responsible for the analysis of form, contour, color, detecting and evaluating objects. The inferior temporal (IT) cortex and associated regions have been implicated in conscious visual perception.

The dorsal pathway moves from V1 through MT and into the posterior parietal (PP) cortex. From there it sends a far-flung projection into the dorsolateral prefrontal cortex. These neurons are concerned with space, motion and depth. It processes visuo-spatial cues necessary for reaching and guiding the eye, hand or arm to appropriate destinations.

Both streams have extensive inter-connections and cross linkages. Some areas, particularly in and around the superior temporal cortex lie at the interface between the two pathways and defy any simple classification.

# Beyond V1

The primary visual cortex represents the world in multiple low and high-resolution maps. These emphasize canonical image features such as orientation, changes in image, wavelength-specific information and local depth. Yet it is but the first cortical area of many discovered to date that subserve vision in mammals. In humans, almost 25% of the entire cerebral cortex is known to mediate visual perception and visuo-motor tasks.

Any accessible region of the cortex can be "turned off" by cooling it with metal plates placed on the surface. When V1 is shut down in this manner, visual responses throughout the so-called ventral hierarchy are much reduced, so much in some areas that even a basic receptor field is compromised.

However, region MT (middle temporal) that is known to mediate motion processing seems to retain some degree of selectivity for movements. MT is mainly fed by two "tributaries", both of which originate in the retina. One passes through V1 while the other reaches the cortex via the superior colliculus (SC). Consistent with this view is the observation that lesioning the corresponding regions in both V1 and the SC eliminates all responses from MT cells.

This cortical bypass may be adequate to support the minimal, unconscious visuo-motor behavior observed in "blindsight" patients whose V1 has been destroyed but yet insufficient to "power" the ventral pathway for conscious, object vision.

V2, the second visual area encircles V1 and is about equal in size. The neurons from V1 project in a one-to-one correspondence to their counterparts in V2, resulting in a similarly skewed topographic representation there. This mapping extends in a continuous manner across the cortical sheet. There are no abrupt borders but the receptive fields get larger and more diffuse.

The V2 neurons are sensitive to depth, motion, color and form. Many are end-stopped, responding best to short bars, lines or edges. No intensity change is present yet one sees contours in the representative visual field defined by either contrast, motion, depth or illusory edges. Its neurons are implicated in identifying partially occluded figures in the visual scene (Where's Waldo?).

Directly adjacent to V2 is a third visual area, V3 with a split mirror-image representation of the visual space, one for the upper and one for the lower field. In front of these are the V3A and V4 areas possessing their own retinotopic map based on inputs from V1, V2 and V3. Its receptive fields are bigger than those of its inputs and the visual information undergoes further evaluation by "fragmentation".

In a series of influential papers, Semir Zeki of University College, London, England suggested that the V4 area mediates color perception in the macaque monkey. Many V4 neurons represent color rather than raw wavelengths of light. However, such color-selective cells are not limited to V4 but present elsewhere in the visual cortex.

In humans, lesions along the ventral surface of the occipital and temporal lobes, part of the fusiform gyrus can selectively disturb color vision i.e. the form and other aspects of vision are present but the hue is gone. This has since been confirmed by fMRI studies done on volunteer subjects.

Intriguingly, some color-tuned regions remain active when subjects experience color after-images in the absence of any physical color. For instance, if you stare for some time at a vivid color (green) and then look at a uniform gray field, you will see its complementary color (red) hang for a short time then fade away. fMRI activity in a section of the fusiform gyrus follows the percept, increasing in response to the virtual color after-image and degrading back to baseline after the inducing color patch has been removed.

In synesthetes, individuals with the unique ability to perceive events in multiple modalities simultaneously i.e. colored-hearing, blind-sight, hearing shapes or seeing touches, hue percepts evoked by words trigger brain activity in the same part of the fusiform gyrus as colored stimuli. Surprisingly, areas V1 and V2 did not show any fMRI response during colored hearing.

# Visual Perception

A central hypothesis in visual perception is that visual information is distributed across distinct subsystems. In this view, perception is analytic. The early processes are devoted to analyzing attributes of a stimulus: Some processes represent shape, while others mediate color and provide information about the dynamics or movement in the visual scene.

In some ways, this may sound almost counterintuitive for our perception of the things out there come across as a unified whole rather than a compilation of visual fragments. Nonetheless, as we shall see later on, converging evidence from various branches of cognitive neuroscience do provide compelling support for the idea that perception operates in an analytic manner.

Indeed, the feature-extraction hypothesis is one of the best examples of how cognitive optometry can provide complementary evidence of the central role played by the visual system in the complex process of cognitive perception. Although each visual area provides a map of external space, the maps differ with regard to the type of information they represent. For instance, neurons in some areas are highly sensitive to color variation. In other areas, the neurons may be movement sensitive but color insensitive.

By this hypothesis, neurons within this area not only code for where an object is located in visual space but also provide information about the object's salient attributes. Visual perception is, in other words, a divide-and-rule strategy. Rather than each visual area representing all attributes of an object, each provides its own limited analysis.

Image processing is distributed and specialized. As we advance through the visual system, different regions elaborate on the initial information gathered by the retina and begin to integrate it to form coherent, recognizable percepts. Extensive physiological evidence supports the specialization hypothesis.

For example, single-cell recordings in the M pathway show that these neurons are not specific in terms of the color of the stimulus. These cells extending from the magnocellular layer of the LGN through 4b and V2 respond similarly whether the colored circle is green or red but very specific to the direction of its motion.

Does visual imagery share space with vision in the brain? Numerous PET scans of subjects show that mental images do tend to be laid out across the visual cortical surface as a sort of crude depiction of the visual field. Interestingly, visual images though they share brain areas with perception are somehow different from the real display. Donald Symons asserts that reactivating a visual experience may well have benefits but it also may confuse imagination with reality.

That may be why within moments of awakening from a dream, our memory for its plot is wiped out, presumably to avoid contaminating autobiographical memory with bizarre confabulations. Similarly, our voluntary, waking mental images might be hobbled to keep them from becoming hallucinations or false memories.

Imagery is a wonderful faculty but we must not be carried away with the idea of pictures in the head. There is no such thing. We cannot reconstruct an image of an entire environment in our heads. We create fragments and collages in our heads. We recall glimpses of parts arrange them in a mental tableau and then sift through the fragments to refresh each one when it begins to fade.

In addition, visual memories are not accurate pictures of whole objects but a sort of caricature or cartoon depiction of the real thing. They cannot serve as adequate models or concepts to cognition or even perception. How can a concrete image represent an abstract concept? For example, how can one conceptualize an American or the idea of freedom or even what constitutes a human being?

Pictures are by definition ambiguous artifacts whereas thoughts cannot be. An image can be worth a thousand words but that is not necessarily such a good thing. At some point between gazing and thinking, images must give way to ideas and perception. Perception is found to be intimately interwoven with memory. Object recognition is more than linking features to form a coherent whole. That "big picture" triggers memories.

Are there separate representational systems for different types of information such as objects and faces? Do the sensory modalities have their own memory systems? Or, do they access a modality-independent knowledge base?

At an even more fundamental level, perception and recognition do not appear to be unitary phenomena but are manifest in many guises. As seen earlier, the pathways carrying visual information from the retina to the first few synapses in the cortex clearly segregate into multiple processing streams.

Early on there is partitioning into the M and P pathways followed by differential projection of the latter to blob and inter-blob zones in V1. But, once past the lower cortical regions, convergence and divergence become the anatomical rules leading to the so-called dorsal (where) pathway of the parietal lobe and the ventral (what) route of the temporal lobe. The neurons in both lobes have large receptive fields.

Neurons in the parietal lobe, however, have an interesting property in that they can respond in a non-selective way (react to both focal and ambient stimuli) in central and peripheral regions of the visual field. The temporal lobe neurons are activated by stimuli that fall within the left or right visual field but always encompassing the fovea. This disproportionate representation of central vision appears to be ideal for a system devoted to object recognition.

Cells within the visual areas of the temporal lobe have a diverse pattern of selectivity. One study employed a wide range of stimuli. Some were simple, involving edges or bars at orientations that varied in color or brightness. Others were complex and included photographs or 3D models of objects like a human head, hand, apple, flower and snake.

Of the 151 cells sampled, 110 consistently responded to at least one of the stimuli. A large minority (41%) acted similar to parietal neurons i.e. they were activated by any of the stimuli and their firing rates were similar across the set of stimuli. The remaining 59% exhibited some selectivity and responded more vigorously when viewing complex stimuli.

Recording from a cell located in the inferior temporal (IT) cortex displayed pronounced activity when viewing a model of the human hand. The activity was high regardless of the hand's orientation and was only slightly reduced when the hand model was considerably shrunk down.

To summarize, the what-where or what-how dichotomy offers a functional account of two computational goals for higher visual processing. This distinction is best viewed as a heuristic one rather than reflecting an absolute distinction. The dorsal and ventral pathways are not isolated from each other but communicate extensively. Processing within the parietal lobe, the termination of the "where" pathway serves many purposes.

It is known to play a critical role in selective attention and the enhancement of processing at some locations instead of others. Moreover, spatial information can also be useful for resolving "what" problems. For instance, depth cues help to segregate a complex scene into its component objects.

# Section 4

# Neural Substrates of Awareness

The neural substrates of awareness or NSA, are by definition a minimum set of neuronal events which when taken together mediate a specific state of awareness (SOA). A specific NSA of, for example, vision accompanies every percept whether conscious or subliminal.

As mentioned earlier, the prevailing SOA is mainly the result of the interplay between three neural systems in humans, one causing arousal and the other two, sleep. The Reticular Formation (RF) mediates all three.

Running through the entire brainstem is a core of tissue called the RF, which is composed of a 'diffuse' collection of small, many-branched neurons (neural net). These neurons receive and integrate information from many sensory pathways as well as from other regions of the brain. Some RF neurons are clustered together forming certain parts of the brainstem nuclei and 'centers'.

The output of the RF divides functionally into ascending and descending systems. The descending components influence both somatic and autonomic motor neurons but frequently, sensory ones as well; the ascending components affect such things as wakefulness and the direction of attention to specific events.

The neural pathways mediating the waking state, arousal and attention lie within the RF and are part of the Reticular Activating System (RAS), destruction of which produces coma and an EEG pattern characteristic of a sleep state. The RAS is crucial not only for the maintenance of a waking state but also for maintaining arousal and attention. It is important to keep in mind that human beings are conscious of a stimulus only when the CNS is oriented and appropriately receptive toward it.

**Diagram** illustrating the neural substrates of Awareness

Tier 3

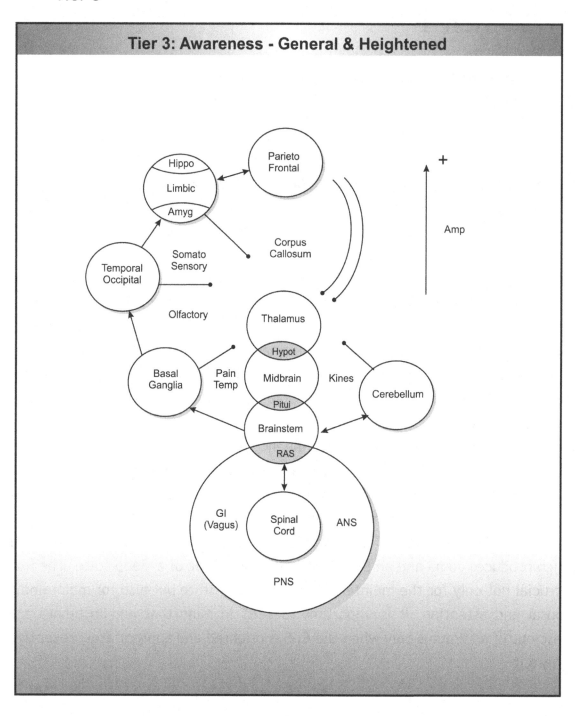

The neural substrates of visual awareness, NSVA for short, comprise of a temporary coalition of neurons that is currently mediating the awareness of a specific event in the outside world. The winning coalition emerges by suppressing competing neural inputs until such time when attention either pulls away from that event or is superseded by a novel stimulus.

Coalitions vary in size and character. For example, consider the difference between actually seeing a scene unfold before you and imagining it later in a contemplative moment. The coalitions required for abstract imagery are likely to be less widespread and far-reaching than those mediating actual seeing.

This is probably because the cortico-cortical feedback connections from the front of the cortex back into the relevant sensory regions are, without help from a sensory input, unable to recruit the large coalition needed, to fully express the sundry aspects of an object or event. They may not reach down (or up) within the cortical processing hierarchy.

For the winning coalition to emerge as a conscious percept or awareness, it may have to cross a neurological threshold, below which it languishes in a sub-conscious realm. How far up the hierarchy the initial net-wave travels depends upon expectation and selective attention.

The linguist Ray Jackendoff's intermediate-level 'theory' of consciousness postulates that the inner world of thoughts and concepts is forever hidden from consciousness, as is the external, physical world, including the body.

One interesting consequence of this hypothesis is that many aspects of high-level cognition, such as decision-making, planning and creativity are beyond the pale of awareness. These operations are, allegedly carried out by a nonconscious homunculus, a highly speculative construct invoked by Koch, which seems to reside in the front of the forebrain. This entity receives streams of information from the sensory regions in the back and relays its output to the motor system for appropriate action.

A further intriguing consequence that follows is that you are not directly conscious of your thoughts. You are conscious of only a re-representation of these in terms of sensory qualities, particularly visual imagery and inner speech.

# Memory & Awareness

Turbellarian worms and planaria, crabs and lobsters, octupuses and squids, houseflies and butterflies do not sleep as we do. They rest for brief periods of time, but this is a very different state from the one we recognize as sleep. Similarly, neither fish nor amphibians display any electrical signs of brain sleep (Hobson 1989). Perhaps in creatures where the unconscious state, the rest period, is not so dramatically different from the awake, moving period, the degree of consciousness may not be very developed and extensive. Indeed, the degree of consciousness might be so vestigial as to be non-existent.

So, coming back to the tube worms... are they conscious at all?

Do we need to decide where in the animal kingdom to draw the line above which they are to be considered "conscious" while those below are mere automata?

Gerald Edelman, a Nobel laureate and biologist, asserts that animals such as a lobster are not conscious and that even so-called conscious animals possess a kind of "primary consciousness" that allows them to just live for the moment and not really count on a memory of past experiences to live.

Edelman sees a liberation from the present as a higher-order consciousness. This condition encompasses an ability for internal representations, which he refers to as the "conceptual symbolic". To him it is clear that only with such higher-order consciousness would we be able to daydream, reminisce or speculate; only with the ability to form subtle concepts would we be aware of ourselves.

Although Edelman concedes that higher-order consciousness may be present in some of the great apes, the operational definition of this type of consciousness does raise nagging questions concerning animal consciousnesses in general i.e. do other animals not have any memory beyond a covert system of values? Are they capable of forming 'internal representations' as found in higher-order consciousness?

Edelman and Tononi consider memory to be a *central* component of the brain mechanisms that mediates consciousness. They question the widespread assumption that memory involves the inscription and storage of information in some kind of coded form. They see it as being *nonrepresentational*, resulting from the *selective* matching that occurs between the ongoing, distributed neural activity and various signals from the world at large. The synaptic alterations that ensue affect the future responses of the individual brain to similar or different signals during recall.

Such a memory has properties that allow perception to alter recall and vice versa. It is robust, dynamic, associative and adaptive with an unlimited capacity. It is not strictly replicative like the one found in computers, which tends to be representational and follows a specified, predetermined code but creative since it generates 'information' by actual construction in real time.

In this view, there are hundreds, if not thousands of separate memory systems in the brain, which can range from all the perceptual systems in different modalities: vision, smell, touch, hearing etc., to ones that govern intended or actual movement, to the multi-variant language systems, which organize the complex sounds of human speech

The whole point of our brain and CNS, from the sensory organs that feed information to the massive networks of neurons that interpret it, is to develop a sense of what is happening currently and what will transpire in the near future. Our brains are basically "prediction machines" and to fulfil their task, they need to find order in the "chaos" of possible memories.

In order to function appropriately, our brain needs to cull through the 'mountains' of extraneous information that we encounter everyday in modern society and extract, screen or filter out the chaff from the grain. And besides, most of this information deluge does not need to be remembered or retained in our consciousness once a decision has been made or an appropriate response recorded.

As we have seen thus far, the brain relies on large populations of neurons acting in concert to represent and form a memory of an organism's life experiences. Neuroscientists are beginning to close in on the "rules" or neural 'code', if you will, that the brain uses to lay down memories. They have discovered subsets of event-mediating neurons, dubbed 'neural cliques' in the mouse hippocampus (an area critical to memory formation) that, seem to respond to different aspects of an event.

For instance, some may help mediate abstract or general information that allow us to transform our daily experiences into a repository of knowledge and ideas while others clue into more specific or selective features of the event like the rumbling of thunder, the sound of shattering glass or the ruffling of a book's pages.

This sort of 'hierarchical' organization could enable the brain to convert such 'streams' of electrical impulses into perception, memory, knowledge and ultimately behavior.

Such an understanding could allow investigators to develop more seamless brain-machine interfaces, design a whole new generation of smart computers and robots or perhaps even assemble a 'codebook' of awareness that would eventually lead to deciphering what another human being remembers and thinks.

## Emotions & Awareness

The belief that emotions are animal legacies from our ancestral apes, snarling and biting their way through conflicts, is a familiar depiction in pop ethology documentaries and the Discovery channel but quite untrue. Most parts of the human body did come from ancient mammals and before them ancient reptiles but the parts are heavily modified to fit human anatomy and physiology.

Emotional repertoires vary wildly among animals depending on species, gender and age. The common chimp often dwell in "gangs" in which belligerent males have been known to massacre rival bands, mount complex monkey hunts and indulge in frank cannibalism while the female sex have been observed to commit infanticide with impunity or snatch another's baby and rip it to shreds.

Their bonobo cousins, on the other hand, profess social alacrity, mount anyone around them sexually and go about living the vida loca till well into adulthood.

There is no question that moods dramatically affect human lives and behavior. Moods and emotions are also critical for survival and

affect the way we perceive things out there. The systems work in tandem, integrated by many reciprocal connections. Most primates are not information-processing machines but rather motivated, emotional and social creatures. They are endowed with one of the most fundamental of human-like characteristics: the ability to feel emotions and express them.

Although there is considerable debate as to whether any single list is adequate to capture the vast nuances of emotional experience displayed in the world at large, most people would accept the idea that there are a finite number of basic, universal human emotions like happiness, sadness, fear, disgust, anger or surprise. These can be easily depicted via a series of facial expressions or reactions to events in the world that vary along a continuum.

One factor that differentiated emotion form other behaviors is that emotion alters not only our mental and neural states but also our physiological state. Emotional responses can bring about a number of bodily reactions because of activating the autonomic nervous system (ANS). For example, a late night jaunt through the alleyways of downtown Manhattan can speed up your heart, raise goose bumps or even induce some anxiety upon hearing quick footsteps behind you.

(**Author's Note**: This idea that emotions alter our physiological state and that we can determine an emotional reaction by measuring our body's response is the basic principle behind the lie-detector test.)

However, is there a clear dividing line between emotion and cognition?

One of the more recent debates addressing this issue occurred between Robert Zajonc of Stanford and Richard Lazarus of UCal, Berkeley. Zajonc argued that affective judgments occur before and independently of cognition whereas Lazarus asserted that emotion could not occur without cognitive appraisal.

The primary issues in this debate hinged largely on how one defined cognition. Zajonc defined it as a slower mental transformation of sensory input or information processing. His studies were conducted to show dissociations between evaluation and awareness.

For example, emotional stimuli were presented subliminally and so quickly, that subjects did not report, "seeing" them but seemed to influence how the subjects evaluated emotionally neutral stimuli that followed.

Lazarus defined cognition as including early evaluative perception as well as later stages of information processing. For example, if we experience signs of ANS arousal such as sweating, increased heart rate and blushing, our emotional response will depend on whether we are working out, having an interlude with a ravishing brunette or looking down from the glass floor of the CN Tower in Toronto.

In other words, the emotional response depends on the reason we believe we are experiencing the arousal. He believed that emotions were a subset of cognitive processes. One benefit of the recent work on the influence of emotions on cognitive processes is that by identifying some of the brain systems of emotion, we have shifted the debate to more neutral grounds.

For instance we now know that some brain structures, like the amygdala, is specialized to process emotional stimuli and can respond very quickly and early in the processing of these stimuli.

These findings seem to be consistent with Zajonc's position that we have separate systems for the processing emotion. At the same time, the neural structures that are specialized for emotion can interact with and be influenced by neural systems known to mediate other cognitive behaviors. These results, however, suggest that emotion and cognition are interdependent, invoking Lazarus's proposal.

The modern infatuation with human emotions and consequent effects on behavior began with the so-called Triune brain theory initially proposed by Paul MacLean, a neuroscientist in the early 1970's. He took the Romantic doctrine and translated it into neuroscience by describing the human brain as an evolutionary palimpsest of three entities:

The reptilian brain as being responsible for primitive and selfish emotions that drive the four F's of feeding, fighting, fleeing and er, sex. Grafted onto it, was the limbic system or primitive mammalian brain, thought to mediate kinder, gentler social emotions like parenting.

And, wrapped around these two was the modern mammalian brain or neocortex, which was believed to house the tumultuous emotional baggage of archaic humans dwelling in the Paleolithic epoch of yesteryears.

After an unbelievable run of nearly four decades, this theory was finally relegated to the recycle bin as neuroscientists realized that natural selection does not just heap layers of cortical matter willy-nilly. It tinkers and modifies the neural substrates of cognition and emotions to fit modern lifestyles.

# The Brainstem

The brainstem is literally the stalk of the brain, through which pass all the nerve fibers relaying signals of afferent input and efferent output between the spinal cord and the higher brain centers. It consists of the medulla, the pons and the mesencephalon or mid-brain. A continuous, fluid-filled cavity, part of the brain's ventricular system, stretches throughout the brainstem. It has two dilated parts: the 4th ventricle is at the level of the Medulla and Pons, whereas the 3rd ventricle lies in the diencephalon.

Examination of its internal structure shows that it is more diverse than the spinal cord. The gray matter is subdivided into several regions or nuclei that mediate common tasks, separated by wispy strands of white matter. It contains the cell bodies of neurons whose axons go out to the periphery to innervate the muscles and glands of the head region. It gives rise to 10 of the 12 pairs of the cranial nerves. The brainstem also receives many afferent fibers from the head and visceral cavities.

Functionally, the brainstem can be looked upon to have two levels of organization. On the one hand, most of the cranial nerves and their nuclei represent "the spinal cord of the head". One the other hand, many cell groups in the brainstem represent a superior level of control over the spinal cord and the cranial nerve nuclei via the so-called Reticular Formation.

# Reticular Formation

In a series of landmark studies in the late 1940's, Moruzzi and Magoun demonstrated that running through the entire brainstem is a core of 'diffuse' nervous tissue, called the Reticular Formation (RF). It is built of cells of various forms and sizes that appear to be randomly mixed. Between the cells, there is a wicker-work of fibers, partly axons and dendrites covering a large volume of neural tissue.

A more detailed analysis reveals that the RF consists of several sub-divisions, among which the cells differ in shape, size and configuration with rather diffuse borders. Its long ascending and descending efferent fibers give off numerous collaterals on their way through the brainstem.

These ensure that signals from each reticular cell may reach many functionally diverse cell groups such as other regions of the RF, cranial nerve nuclei, dorsal column nuclei, colliculi, spinal cord, cerebellum, certain thalamic nuclei and the hypothalamus.

The ascending fibers of the RF end in the intra-laminar thalamic nuclei unlike the specific sensory tracts that end in the lateral thalamic nucleus. These are of particular importance for the general level of activity of the cerebral cortex, which seems to mediate consciousness and attention or the level of arousal or wakefulness in animals.

It was noticed that direct electrical stimulation of this multi-faceted structure via microelectrodes, seemed to 'awaken' the forebrain. The EEG changed abruptly from the slow, high amplitude, synchronized waveforms characteristic of deep sleep to the fast, low amplitude, desynchronized activity typically found in the awakened animal.

# The Reticular Activating System (RAS)

This notion of a 'monolithic' activating system has given way currently to the realization that 40 or more highly heterogeneous nuclei with novel cell structures are housed within the brainstem. The basic cyto-architecture of each cellular cluster is extremely 'amorphous' in stark contrast to the linear, layered organization of the cerebral cortex.

The cells in these clusters manufacture, store and release numerous neuro-modulating agents such as acetylcholine, serotonin, dopamine, norepinephrine, histamine etc. These receive and integrate information from many afferent pathways as well as from many cortical areas of both cerebral hemispheres. Essentially, the RF attends to tasks involving complex behaviors, which include the control of body posture, orientation of the head and body toward external stimuli, control of eye movements etc.

A considerable number of spinoreticular fibers project to widespread regions of the brainstem reticular formation. These fibers are predominantly uncrossed and terminate chiefly upon cells of the nucleus gigantocellularis of the medulla. The spinoreticular fibers passing to pontine levels are distributed bilaterally in the pontine reticular nuclei.

A small number of spinoreticular fibers reach the midbrain RF. Functionally, the spinoreticular fibers play a significant role in behavioral awareness and in the modulation of electrocortical activities.

Two relatively large regions of the brainstem RF give rise to fibers that descend to spinal levels. One is in the pontine tegmentum, while the other lies in the medulla.

Experimental studies indicate that stimulation of the brainstem RF can:

(1)  Facilitate and inhibit voluntary movement, cortically induced movement and reflex activity,

(2)  Influence muscle tone, probably via the gamma system,

(3)  Affect phasic activities associated with respiration,

(4)  Exert pressor and depressor effects upon the circulatory system, and

(5)  Exert facilitating and inhibiting influences upon the central transmission of sensory impulses.

Areas of the medullary RF from which the medullary reticulospinal tract arises correspond to regions from which inspiratory, inhibitory and

depressor effects are elicited. The facilitory, expiratory and pressor effects are obtained from regions rostral to the medulla.

The brainstem RF also receives inputs via cortico-reticular projections from widespread areas of the cerebral cortex, although the greatest numbers originate from the 'motor area'. The regions of termination correspond to those that give rise to the reticulo-spinal tracts. Thus, the synaptic linkages of cortico-reticular and reticulo-spinal fibers form a pathway from the cortex to spinal levels. There is no evidence of a somatotopic arrangement within this system.

Finally, most of the autonomic pathways originating in or relayed via hypothalamic neurons, project to the autonomic centers in the brainstem tegmentum, which relay the impulses to the various spinal neurons. It is striking that the efferent connections of the RF can activate both lower and higher levels of the CNS.

Electrical stimulation leads to alteration of several functions mediated by the spinal cord, such as muscle tone, respiration and blood pressure. Also, the general activity of the cerebral cortex, closely related to the level of consciousness

# Some Specific BrainStem Nuclei

Locus Coeruleus and the Raphe

Near the central gray of the upper part of the 4$^{th}$ ventricle and under its floor, lies an irregular but compact 'mass' of approximately 15,000 deeply pigmented neurons known as the Locus Coeruleus (LC). It appears to be present in all mammalian species. Fluorescein studies have revealed that the cells of this region contain catecholamines, practically all of which are norepinephrine.

Such adrenergic pathways, originating from the LC ascend to diencephalic levels, distributing widely in the cerebral cortex, the hypothalamus, the basal ganglia and the hippocampal formation, some even terminating in the cerebellum. Their organization appears to influence neural activity simultaneously in broad cortical regions of the brain. These neurons respond preferentially to novel 'exciting' stimuli and play an important role in mediating 'arousal', wakefulness and the control of sleep.

The Raphe nuclei form a narrow strip of neurons running along the midline of the medulla, pons and midbrain. They contain cell groups that technically belong to the Reticular Formation (RF), but appear to serve distinctive functions. Together, the raphe nuclei receive afferents from the cerebral cortex, hypothalamus and the RF while the efferents travel south to the spinal cord and north towards the cerebellum, PAG, hypothalamus, thalamus, hippocampus, amygdala, regions of the striatum and several septal nuclei.

Histofluorescent studies demonstrate that these cells contain stores of serotonin and other monoamines. Both serotonin and norepinephrine-containing neurons in the RF play active roles in the mechanisms that control sleep states, moods and emotions.

Inhibition of serotonin synthesis or the total destruction of the serotonin-containing neurons in the Raphe system leads to total insomnia. Serotonin appears to be involved in the neural mechanism related to so-called slow wave sleep. In addition, serotonin seems to modulate the neurons of the locus coeruleus region, which trigger REM or paradoxical sleep.

A final peculiarity of the raphe nuclei is that they send fibers to the ependymal cells, which line the inner aspects of the brain ventricles. Their function remains unknown but the presence of serotonin raises speculation of homeostatic control mechanisms.

# Neuro-modulators

Acetylcholine, dopamine, norepinephrine and serotinin are members of the class of brain chemicals called neuromodulators. Although they do not themselves convey signals about content, they alter the form or mode of how the brain ends up processing this content i.e. 'priming-the-brain-pump' to react/respond in a certain way.

Pharmacology and cognitive sciences are thus in a three-way interaction with physiology that determines the 'form' of conscious experience. That form shapes and constrains its content. And much of this process appears to be due to chance.

The basic rule of psycho-pharmacology is that many of the drugs currently prescribed by psychiatrists and other physicians treating patients for neurological problems act on the brain-mind state control systems of the brainstem. What this essentially means is that the most potent legal drugs (medications) prescribed today share common mechanisms with the illegal street ones.

A sobering corollary of this principle is that the differences between legal and illegal drugs are never as sharp as the authorities who make and enforce the laws would have us believe. By definition, nature is quite economical in what it designs via the culling process of natural selection. It uses many of the same chemicals to accomplish tasks not only within the CNS but throughout the body.

For instance, the REM sleep modulator, acetylcholine, within the Pons, tends to promote sleep, whereas in the midbrain and medulla it promotes alertness. How come? Because acetylcholine is not acting by itself in both these situations. There are other chemical modulators, like serotonin, norepinephrine and possibly a mix of others that mediate your personal sleep-wake cycle.

The neuromodulatory systems of the brainstem directly alter normal SOCs such that we experience psychosis-like phenomena in our dreams. Drugs that increase or decrease the efficacy of any one of these systems are therefore likely to be quite effective in achieving desirable shifts in the cognitive and emotional symptoms of spontaneous altered states like schizophrenia or depression.

These benefits may come at a price, however, as other "state" components are always tied to cognition and emotion at a deep mechanistic level. The sad conclusion is that the medical profession and the pharmaceutical industry may be collaborating in an unwitting and unplanned program of experimental medicine.

For those physician-scientists like us at the interface of cognitive neurology that are interested in studying consciousness, the opportunities presented for understanding are strongly countered by the ethical dilemmas raised. Does that mean that in the interests of consistency we should make all drugs legal?

The political, social and ethical implication of this query are enormous. But in order to develop a balanced and informed approach, we need to look more closely at various aspects of clinical pharmacology.

And, of course, in keeping with the tenets of territoriality and protection of the proverbial "turf", each scientific or professional discipline should look deep within and arrive at some rational choices.

# Neuro-Peptides

There is a large class of chemicals that exceeds all the other bioactive transmitters in terms of variety. These chemicals have a classic signaling action yet can also function as hormones, thereby accessing the endocrine and immune systems as well as the nervous system. The peptides thus link all the major control systems of the body.

Peptides are larger molecules than the better-known transmitters such as acetylcholine and the amines. As far as brain function goes, they have long proved a puzzle in that they are stored with classic transmitters in the same neurons, yet have a subtly different role.

They tend to be released into the synaptic cleft only when the firing rate of the axon becomes particularly vigorous and the so-called "ripples" in the brain-pond begin to encompass larger and larger neuronal assemblies. The larger the assembly, the more of a particular type of peptide will be released. The greater the turn-over of assemblies, the more variants of small amounts of peptides there will be at any one moment.

This readout will signal not only the size of the prevailing assembly, but will also convey information regarding its content, reflected in various levels of different combinations of different peptides, released from different brain sites, determined by a varying rate of turn-over.

Peptides, therefore, are the perfect intermediaries between net brain states (net neuron assembly size and turn-over rate), the endocrine and immune systems and the vital organs. In line with this idea, some researchers have recently reported that agents blocking the action of a certain peptide, Substance P, actually have a novel, anti-depressant action.

Over the last two decades, one researcher has gone on to show just how pervasive and flexible the peptides can be orchestrating one's emotions and indeed, health. In fact, some have actually claimed that peptides are not only the "molecules of emotion" but that there may well be a different peptide for each emotion!

# The Limbic System: A modern approach

On the medial surface of the cerebral hemispheres on either side lies a large arcuate convolution formed primarily by the cingulate and para-hippocampal gyri. Collectively known as the '*grande lobe limbique*', these surround the rostral brainstem, the inter-hemispheric commissures, the hippocampal formation and the dentate gyrus.

From a phylogenetic and cyto-architectural point of view, the limbic lobe consists of the allocortex (hippocampal formation and dentate gyrus), the paleocortex (pyriform cortex of the anterior parahippocampal gyrus) and the mesocortex (cingulate gyrus). The striking feature of the limbic lobe is that it appears early in phylogenesis and possesses a certain constancy in gross and microscopic structure.

The designation of the limbic system stands for an even more extensive and inclusive term, used to include all of the limbic lobe and its associated subcortical nuclei i.e. the amygdaloid complex, the orbitofrontal cortex, the septal nuclei, hypothalamus, epithalamus and various thalamic nuclei. The medial tegmental region of the midbrain is also included since it contains both the ascending and descending pathways subserving the hippocampal formation and the amygdaloid nuclear complex.

Despite the heterogeneity and diffuse nature of the so-called limbic system, there are compelling observations that the structures comprising it are involved in neural circuitry that gives rise to a 'subcortical continuum'. It begins in the septal area and seems to extend to the paramedian zone through the preoptic region and then via the hypothalamus into the rostral mesencephalon.

The intimate relationship of the limbic lobe with the hypothalamus has led some neuroscientists to regard the limbic system as a whole to function as 'a visceral brain'. In humans, it seems to occupy a central position in mediating behavior and the complex world of emotions.

Over the years, the investigation of emotion has intensified. Researchers now acknowledge that emotion, especially in humans, is a multi-faceted behavior that is not amenable to a simple definition or mediation via a single neural circuit or cortical system. Studies of the cognitive neuroscience of emotion have invoked a number of brain regions with the orbitofrontal cortex and the amygdala emerging as prime suspects.

As we gain a greater understanding of the relative roles of these neural structures in emotional processing, it has become more apparent that we need to understand how these neural systems and others interact to produce normal and adaptive emotional responses.

For instance, most of the immediate focus on generating and comprehending the spoken words takes place in the prefrontal lobes within milliseconds, while their emotional content is mediated by the amygdala via the pituitary gland over the next few minutes.

This apparent mismatch in 'processing' gives 'meaning' and context to the scene i.e. Qualia. The amygdala largely operates below conscious awareness and regulates autonomic behavior that we cannot directly control.

Perhaps the most important insight to come out of the growing understanding of our brain's chemistry is what the experts call 'mood congruity' i.e. the brain tends to record not only the specific details of the event but also our feelings about it.

Our memory system tends to serve up recollections of past events that are themselves 'congruous' or similar in emotional content to your current mood. Our emotional state skews our sense of perspective by seeking out memories that match our current mind-set instead of a balanced, representative sample.

Emotions do not merely mark certain memories as being more important than others are. They also affect which details are recorded. Images that trigger strong emotional responses are recalled more readily than neutral ones. In addition, interestingly, so-called 'happy' images leave behind more of a generalized feeling of pleasantness while negative ones are recalled in excruciating detail.

We are also 'wired' to remember novelty and events that somehow deviate from expectation. In fact, we tend to dwell on events that take us by surprise. Researchers now believe that there is an entire neurochemical system devoted to the pursuit and recognition of new experiences and surprise, particularly ones pertaining to reward.

It is regulated by Dopamine, one of the brain's 'reward' drugs.

If the pleasure factor of a certain event exceeds expectations than there is a spike in dopamine presence within the system, whereas disappointment causes a dramatic reduction. It is also believed that chronically low levels of dopamine play an important role in 'addictive' behaviors.

The last among the limbic structures we shall address are the hippocampus and adjoining regions of the temporal lobe, the dentate gyrus, the subiculum and the entorhinal area of the parahippocampal gyrus. Collectively these form the hippocampal formation (HF), which is precisely organized with several different cell types interconnecting in highly complex patterns.

Two aspects of these connections are central to understand its functional roles in mediating cognition: first, the extensive reciprocal connections with numerous cortical association areas and second, the direct and indirect links with other limbic structures (amygdala, cingulate gyrus and septal nuclei).

As for the neocortical connections, the hippocampus obviously processes large amounts of information. The parallel increase in its size during evolution furthermore indicates that its main functions are related to the neocortex. The quantitatively dominating afferent inputs to the dentate gyrus and the hippocampus arise in the entorhinal area. This area receives afferents from nearby association areas (the parahippocampal and perirhinal cortices) of the temporal lobe and integrates various kinds of sensory information from the cingulate gyrus, the insula and the prefrontal cortex.

The neurons of the dentate gyrus send their efferents to the hippocampus, whereas the hippocampal neurons project to the subiculum. From here, the fibers travel through the fornix to the mammillary nucleus and back to the entorhinal area.

A large number of commissural fibers also connect the hippocampi of either lobe indicating a close cooperation between them. The main flow of information from the hippocampus via the subiculum is directed toward association areas in the temporal and frontal lobes.

There is now much evidence that the HF plays a key role in certain kinds of learning and memory. Lesions restricted to the hippocampus proper produce amnesia i.e. impaired memory without perceptible reductions in intelligence that is "shallower" than one that results from damage to the entire HF. The HF appears to mediate short-term memory as isolated damage to it does not seem to erase older ones from a few days ago although it prevents the learning of new material.

Memories can be roughly distinguished into two kinds: declarative or explicit, which relates to the memory of events and facts, while the other called non-declarative or implicit, has to do with the development of skills and habits.

It has been found that only the *declarative* type depends critically on the integrity of the medial temporal lobe. For non-declarative memory, the basal ganglia, cerebellum and regions of the neocortex seem to be more important.

# Section 5

# Animal 'Minds'

Most of us love our pets and cannot understand how people, especially well educated ones, still labor under the delusion that "animals are Cartesian machines and it is the availability of language that confers on us, first, the ability to be self-conscious and second the ability to feel". According to Macphail, there is no convincing evidence for enhanced awareness in other species. They are not just devoid of speech and self-awareness but feelings too.

On the other hand, Baars argues that there are no fundamental biological differences that could justify denying subjectivity to other species for the known correlates of enhanced awareness are phylogenetically ancient, going back to at least the early mammals.

From the observer's point of view, vision feels like an automatic process that helps map out the external, physical reality into the inner, mental universe. However, a few moments of introspection reveal that the relationship between these two worlds is far more complex. Experiences are not simply 'given', as some empiricists have asserted. Rather, your 'mind', implicitly or explicitly, selects the few nuggets of information that are of current relevance from the vast flood of data streaming in from the sensory periphery.

The prefrontal cortex is incapable of processing all this information. It deals with this information overload by selectively attending to a miniscule portion of it, neglecting most of the rest. By selectively attending to particular events or things out there, you choose to experience just one of the innumerable worlds out there clamoring for attention. What you are conscious of is usually what you attend to. Indeed, a venerable tradition in psychology equates enhanced awareness of an object or event with attending to it.

However, it is important not to conflate these two notions. Attention and cognition are distinct processes and their relationship may be more intricate than conventionally envisioned. Some type of attentional selection is probably necessary but may not be sufficient for cognitive perception. When attending to something, the rest of the scene outside your 'focus' still goes on. Even when 'spacing-out' while driving or thinking, one does remain conscious of the gist of the scene in front of you.

Change blindness, where the observer fails to detect a large change between two otherwise identical images, inattentional blindness in which the magician performs obvious sleight-of-hand tricks right under one's nose are just two instances where some type of attentional selection or deficits occur.

Analogous to ambient vision, where any movement out in the peripheral field instantly captures the gazer's interest and compels foveation of the object, neuronal representation for gist perception mediates the rich sense of seeing or in this case, attending, to everything.

# Adaptation and the Brain

Evolutionists researching human behavior think that the modern human brain was adapted to deal with the world of the Pleistocene epoch, about 100,000 years ago. Therefore, when we think about the functions of the modern brain and what it does and does not do well, we should take into account what the early hunter-gatherers had to cope with.

Various studies of decision-making capacities in humans, argue that the mechanisms for decision- making evolved in a particular social context when humans lived for the most part in small groups. There may well be differences in how we make decisions in different social contexts, some tending to appear more irrational than others do. According to evolutionary thinking, these special capacities grow from separate and individual adaptations.

The cognitive system that evolved is thus not a 'unified' system that can work by applying specially crafted solutions to unique individual problems. The adaptations built into our brains are the neural networks, which set us apart from the rest of our great ape cousins. For example, the brainstem, which is the oldest part of the brain (400 million years ago) still, manages to retain, even to this day, many of the anatomical and functional characteristics. This is where we will begin our search for the initial birth of 'true' human enhanced awareness.

The choice of animal that researchers use for their experiments depends on their aim. If researchers are interested in when the MT (medial temporal) visual area, a region of the visual cortex thought to mediate motion detection in early primates such as lemurs, lorises, galagos etc.

Kaas has addressed this question by determining that all of these early primates possess such an area in their brains. We know now through various MRI and computer assisted hardware that numerous 'modules' are progressively added on to these primary sensory regions as the animal rapidly adapts to changing climatic conditions in its environment.

Interestingly a comparative analysis of several non-human mammalian brains to make inferences about so-called 'higher order' association areas, for instance, the pre-frontal cortex, reveals some surprises. Human prefrontal cortex, even though larger in size than that of a macaque monkey, seemed to be 'dwarfed' when compared to a scaled model of the spiny ant-eater, hardly a 'mental' giant.

To the majority of us who tend to view the pre-frontal cortex as the key brain area mediating human uniqueness in the natural world, this result seems odd. However, an evolutionary perspective suggests that the pre-frontal cortex arose from an ancient olfactory system.

In humans, it was integrated into the expanding sensory cortex with a series of multi-synaptic networks. Although there may be some overlapping functions of the pre-frontal cortex in numerous mammal and primate species, this region has been co-opted for more varied functions in the human being.

Another example is Broca's area. A comparative analysis suggests that the most parsimonious and biologically compatible explanation is that the modification to this region of the human brain has homologous counterparts in other primates.

However, the formation of new patterns of neural connections and the clear observed behavioral differences, indicate that it is not analogous. In other words, in humans this region of the cortex was recruited to sub-serve language acquisition.

# Basic Instincts and "Intelligence"

Complex neural circuitry has evolved in many animals but the commonly perceived image of animals following some sort of mental hierarchy is meaningless. This view proposes that the ability to associate cause/effect scenarios gets better as the organism 'rises' higher on the phylogenetic scale.

Eventually, the creature is freed from bodily drives dependent on physical stimuli to respond and is now able to communicate abstract ideas directly via the spoken word. However, the distribution of intelligence in the real animal world does not bear out this conception.

Many species are able to compute how much time to forage at each patch in order to optimize their rate of return of calories per energy expended. Some birds can scan the path of the sun above the horizon and circum-locate their home base. Certain owl species can utilize microsecond disparities between echoes bouncing off surrounding objects and home in on a prey item in pitch darkness.

Rodent species in the habit of caching nuts and seeds for later consumption are able to recall almost all the burial spots spread over a few square acres, a month later. What these examples signify is that animal brains are extremely complex entities and precision instruments enabling it to use information to resolve issues presented by its lifestyle or ecological niche.

Various IQ criteria cannot in all fairness, rank disparate animal species or by the percentage of human intelligence, they are thought to have achieved. Whatever is special about human sentience cannot be an extrapolation signifying more, better or broader aspects of animal intelligence because there is no such thing as generic animal intelligence. Just as an automobile is different from a horse-drawn carriage even though both are used as a vehicle of transportation, each creature has evolved its own unique information-processing machinery to resolve its special set of problems.

Mammalian brains do seem to follow a general plan and the major visible differences emanate from enlarging or shrinking various sections or lobes. The number of cortical areas differs widely from twenty or fewer in rats and mice to upwards of fifty in humans. Primates differ from other mammals in the number of visual areas, their interconnections and their hookup to the motor and decision-making regions of the frontal lobes.

When a species has a noteworthy talent, it is reflected in the gross anatomy of its brain, sometimes in ways visible to the naked eye. The takeover of simian brains by expansive visual areas that tend to occupy almost 50 percent of processing ability reveals their aptitude for locomoting through dense tree canopies.

Bats that rely heavily on echolocation and ultrasound have additional cortical areas dedicated to auditory stimuli processing. Gerbils that cache seeds in numerous locations around the neighborhood are born with larger hippocampi, long considered by neuroscientists to be the seat of cognitive maps and GPS constructs.

A quick look at the human brain shows that the primate brain has been considerably re-engineered to be housed in the spacious calvarium. The olfactory apparatus, hugely prominent in lesser primates has somehow shriveled, the auditory lobes have blossomed into large association areas for understanding speech and the prefrontal areas believed to mediate deliberate thought and planning have ballooned to twice what an average primate should possess.

Even though it is interesting to observe these differences in a gross perspective, the real reason why humans tend to be sentient is in the quality of interconnections among its various regions. Analogous to the high-speed microchips of modern-day computers, the functioning micro-circuitry of the human brain is millennia ahead of the rest of the primates.

Our vaunted flexibility stems from scores of instincts assembled into programs and pitted in intense competitions within human societies. When all goes well, our reasoning instincts are capable of abstract analysis directed towards achieving goals. As humans, we have rediscovered traits that are rare among primates but found in other animals. We walk on two legs like large birds, we live longer than any other great ape and we hunt with impunity, eating meat as a primary source of nourishment.

All organisms evolve defenses against being hunted or eaten setting in motion an evolutionary arms race, if you will. In any generation, the tools or weapons derived to facilitate this condition do not change very much. Experiential knowledge is used to attain goals in the face of obstacles.

Humans are different in that they compose new knowledge and plans by mentally rehearsing hunting or evasive strategies in their mind's eye. Living by their wits, human groups develop sophisticated techniques for foraging and interacting appropriately in a complex social milieu. Language evolved as a means of exchanging knowledge among disparate human groups. An information-exploiting lifestyle goes well with living in groups and pooling expertise i.e. with culture.

Cultures differ from one another because they pool bodies of expertise fashioned at different times and places. A prolonged childhood serves as an apprenticeship of sorts for acquiring knowledge and a useful skill-set. Humans live long enough to repay the investment in a prolonged apprenticeship.

New habitats can be colonized fairly easily because even if the local conditions differ, they obey the universal laws of physics and chemistry that have been acquired from the hard-won wisdom, strokes of genius and the trials-n-errors of preceding human populations. These, in turn, can then be enhanced and refined via appropriate modifications to prevailing conditions in the current ecological niche.

# "Ghost" in the machine?

Once our faculties for language and social interaction are up and running, some kinds of learning may consist of simply recording information for future use, like the name of a person, her address, hobbies and interests or the members of her family. Others may be more like setting a dial, flipping a switch or deciphering social cues, where the 'apparatus' is in place but a 'parameter' is intentionally left open so the 'mind' can track variations in sensory information flow.

This so-called 'mind' may come with a battery of emotions, drives and faculties for reasoning and communicating with a common 'logic' across diverse human cultures. These may owe some of their basic design to genetic information already embedded into the genome, honed or finessed by the culling process of natural selection over the course of human evolution and thus difficult to erase or redesign from scratch.

However, sequencing of the human genome revealed a surprise. Humans, with our superior intelligence and humongous brain sizes possess only twice as many genes as a humble earthworm. Most biologists, of course, who ponder this puzzle, do not conclude that humans are less complex entities. They conclude that genes specify the production of various proteins, some of which may subserve a regulatory function.

Depending on how such genes interact, the assembly process can be much more intricate for one organism than for another with a similar number of genes. It has little to do with the actual complexity of the creature. Heck, mice and men not only have similar genes but essentially very close in number: 23000 in mice to 25000 for humans. Walt Disney was right after all. Mickey is almost human.

Neural network modelers have begun to show how the building blocks of mental computation, such as storing and retrieving a pattern, can be implemented in neural circuitry. Educated people, of course, know that perception, awareness, language and emotions are rooted in the brain. But it is still tempting to think of the brain as a control panel with gauges and levers operated by a 'wizard-like homunculus' usually known as the self, the soul, the ghost, the 'me'.

However, cognitive neuroscience is showing that such a so-called self is nothing but a neural network of brain 'modules'. Evolutionary biology has already revealed that complex adaptations are ubiquitous in the natural world and that natural selection is fully adept at evolving them. Linguists have also discovered that distinct use of symbols and abstractions utilized in various modes of human information transfer either verbal or eye signaling etc. have neural correlates in the brain cortex.

Developmental psychology has also shown, as seen in the previous section on perception and cognitive development that these distinct modes of interpreting experience are initiated early in infancy and get more refined as the child matures.

So we finally arrive at the 64-dollar question: Is there someone home? Does the human 'machine' indeed harbor a 'ghost'?

I will be tackling this profound query in the last few chapters of this book in the human enhanced awareness section.

# Human "Intelligence"

A human being's awareness of the world is mediated by numerous physiological mechanisms, involved in the processing and organization of afferent or sensory stimuli impinging upon the central nervous system. Initially, the incoming stimulus energy evokes *action potentials* within individual nerve fibers.

The action potential has characteristics determined only by the properties of the neuron, independent of the characteristics of the exciting stimulus. It is in a coded form. The code represents information from the external world, differing vastly from the actual incoming stimulus it seems to be representing. For example, the eye does not really form an inverted image of the outside world on the retina as most people believe.

Instead, myriad receptors in the ten layers of the retina break down the visual cues into a steady stream of coded electrical impulses. These flow through the million-odd neurons of each optic nerve to the thalamus (LGN) for further processing. This *edited* information then passes on to the primary visual cortex layers where it is organized into bits and pieces of a complex puzzle.

A database of memories and prior experiences are meticulously accessed via a massive network of reciprocating 'association' fibers that infuse the 'picture' with emotions and feelings. All this information now flows into the 'executive' branch of the brain, which helps us 'register' the scene playing out in the outside world and grant it 'reawareness'. Only 500 milliseconds have transpired between first setting eyes on an object and reawareness!

Afferent information may or may not have a conscious correlate i.e. it may or may not give rise to a conscious awareness of the physical world. Afferent information that does not have a conscious correlate is called sensory information and the conscious experience of objects and events of the external world, which we acquire from the neural processing of sensory information, is called perception.

Intuitively it may seem that the sensory systems of our body operate like some our more familiar electrical equipment used in interpersonal communications, for example, a telephone. However, there are some essential differences. A phone changes sound waves into electrical impulses and back into sound waves.

Our hearing system changes the sound waves into *graded* action potentials, which are processed by the brain as *sound*. The brain does not physically translate the code into sound waves. Currently, just as in the visual system example above, we have no clear understanding of how these coded action potentials are perceived as conscious sensations.

Information about the external world and internal body environment exists in different energy forms e.g. pressure, temperature, light, sound, smell etc., but only specific receptors can deal with them. Even though any one receptor is usually more responsive to one specific stimulus modality, several different energy forms can activate virtually all of them, if the stimulus intensity is sufficient, or if the receptor is unusually sensitive.

The rest of the nervous system can extract meaning only from the graded or action potential in which it is encoded. Several different kinds of information (stimulus modality, intensity and localization) are conveyed in code to the CNS.

All incoming afferent information is subject to extensive control before it reaches the higher levels of the CNS and much of it may be reduced or even abolished by inhibitory input from other neurons. Our actual perception of the events around us often involves areas of the brain other than the primary sensory cortex. Systems for controlling attention include the parietal lobe, temporal cortex, frontal cortex and various subcortical structures.

Information from the primary areas achieves further elaboration through the neural activity of nerve fibers relaying information from cortical association areas. These perform higher mental functions like providing contextual information memory, recall of past circumstantial information and collating it into a more 'personalized' version of the event. All of us put great trust in our sensory-perceptual processes despite the inevitable modification and editing we know that tends to exist in the CNS.

Some of the factors known to distort our perceptions of the real world can be:

Receptor-adaptation to certain recurrent stimuli, either due to receptor fatigue or personal disregard.

Or, confounding factors such as emotions, personality and social background, influencing perception,

Or, not all information entering the CNS is giving rise to conscious sensations.

Or, incompleteness of the real picture due to the lack of suitable receptors for many regions of the electromagnetic spectrum.

Thus, it appears obvious that the twin processes of transmitting data through the nervous system and interpreting it, cannot be separated. The information is processed at each synaptic level of the afferent pathways and there is no one point along its sojourn when it can be interpreted as becoming suddenly 'conscious'.

The apparent limitless-ness of human intelligence stems from the power of a combinatorial system. For example, just as the 26 letters of the English language can be combined into the Encyclopedia Brittanica, a few ideas can combine into a successful space program, capable of transporting human artifacts to the edges of our solar system.

In other words, the ability to conceive an unlimited combination and permutation of mundane ideas is the powerhouse of human intelligence and a key to our success as a dominant species here on earth.

# Somatic Marker Hypothesis

Antonio R. Damasio, a neuroscientist at the University of Southern California has introduced the notion that your feelings strongly contribute to even the most "rational" decision-making in every day life. Most scientists used to assume that reason and emotion were qualitatively different psychic spheres. Clearly these spheres could influence each other, yet the thinking was that the 'cognitive' knowing and reasoning entity of the human 'mind' was in some fundamental way distinct from the feeling, emotional one.

Actually, the idea that we sense our emotions from our bodies (enteroception) has been around for more than a century. Two psychologists, William James and Carl Georg Lange from way back when had hypothesized that emotions arose when you perceived changes in your body.

For example, when accosted by an assailant in a dark alley, you are afraid not because of your rational assessment that you are about to be attacked but because your heart is racing, your stomach and sphincter are clenched and you are mentally searching for a safe way out of this potentially life-threatening situation.

Numerous brain scan techniques brought in to monitor such intense emotional activity reveal significant neural discharge in several cortical regions notably the insula and anterior cingulate. Both these regions are crucial centers of emotional awareness and also necessary for attending to 'feelings' that arise from deep within the body.

One region, the right frontal insula, displayed the greatest activity in subjects that scored highest on a standardized questionnaire designed to probe their 'empathy' levels i.e. the more viscerally aware the person was, the more emotionally attuned they were. Interestingly, in a follow-up study, the scientists found that people with greater empathy have more gray matter in their right frontal insula region. That is, the thicker this portion of the insula, the better you are at reading 'feelings' in yourself and others, lending further credence to the fact that some people are more emotionally aware than others has a neural, physical basis.

Such experiments provide us with a unique 'window' into some of our most important and fascinating body maps of somatic sensation that is oriented inward.

These interoceptive maps in humans are an enhanced version of neural circuitry that had already become highly advanced in primates.

In 'lower' vertebrates such as amphibians, for instance, sensory information is integrated in the brainstem as they do not have a cortex. The cortex imbues the mammalian 'mind' with the capacity to form highly detailed and versatile representations of sights, sounds and planned actions.

However, in all other mammals aside from primates, this homeostatic information from the body does not form a rich interoceptive map in the insula. Because of this difference in mapping, some experts claim that the sensory experiences of cats, dogs, rats or horses, to name a few mammalian species, must be profoundly different from ours, even though we are often tempted to attribute human-like emotions and intentions to our pets.

Whereas a dog may show "shame" through its body language, it does not feel what you feel when you are ashamed. Dogs are clearly emotional and self-aware, but they are not in the same league as humans. In primates, interoceptive information is elaborated via a rich set of mappings in the insular cortex. In humans, it is richer still.

Thus, we have a small insular map for sharp pain, another for burning pain, one for itching, aching, over-exerted muscles and so on, along with several visceral *homunculi* closely monitoring the state of your lungs, heart, GI tract, kidneys etc.

After reading off the internal state of the body from both the right and left insulae, the human brain performs yet another level of integration. This sensory-rich information is routed to the right frontal insula, where conscious physical sensation and conscious emotional awareness co-emerge. This region seems to be the focal point connecting the state of your body to the state of your brain.

This is a profoundly important insight for it is here that the 'mind' and body unite, laying the foundation, if you will, for emotional intelligence in human beings. The right frontal insula also integrates your 'mind' and body through reciprocal connections with three other brain regions: the amygdala, the orbito-frontal cortex and the anterior cingulate.

In every brain-imaging study ever done of every human emotion, the right frontal insula and anterior cingulate cortex seem to 'light-up' together. Some scientists takes this to mean that in humans, at least, emotions, feelings, motivations, ideas and intentions are combined to a unique degree. Complex human emotions, it seems, are never truly divorced from decision-making, even when they are channeled aside by an effort of will.

# Hemispheric Lateralization & Human Uniqueness

From the perspective of natural selection, by which organisms acquire specific adaptations, one would assume that the two cerebral hemispheres would not function identically. After all, one does not need two speech centers or two places to store a memory of the same face. Once a region of the brain has evolved a functional specialization, there would seem to be no need to duplicate it in another region.

However, the cerebral organization suggests that duplication rather than unilateral specialization is the rule. The two hemispheres are much more similar to each other in function than one may think. The differences only surface at a more subtle level of analysis. Each hemisphere's visual areas are devoted to representing the shapes and colors of objects and their primary specialization relates to the object's position: Is the object on the left side of space or the right?

What is more, the motor cortices are roughly mirror images of each other, both in structure and function. While each cortex can be represented by a motor homunculus of the body, it is dominated by fibers that project to muscles on the opposite side of the body. Evolution could have utilized the available cerebral space in a completely different way.

Visual functions could have been isolated in the left hemisphere and motor projections to both limbs could have originated in the right. Yet, selective pressures appear to have favored a cerebral organization that reflects structural properties of the world and the organism. The world's spatial structure is reflected in our biology.

The hemispheric specializations are best conceived as superimposed on this fundamental symmetry. In some instances, specializations may have evolved because there was an advantage in having a single system devoted to a certain process. For example, one hypothesis postulates that speech production became strongly lateralized because of the need to communicate at rapid rates. Transcortical processing and integration take time and may slow down complicated articulatory gestures. Indeed, lateralization may underlie stuttering because of the two parallel systems that compete for control of speech output.

Others argue that hemispheric specializations evolved because of the inherent advantages in having nonidentical forms of representation. Homologous visual areas perform related operations but differently enough that the resultant nonidentical representations are imbued with unique advantages in performing certain tasks.

This does not mean that these tasks are strictly localized, that language functions are restricted to the left or spatial ones to the right. Not only does the normal performance of these tasks require distributed operations that may span both hemispheres but also usually, both hemispheres contain the essential machinery for performing the task.

This organization help explain why patients with large unilateral lesions that damage 25% of the cortical tissue on one side still display an amazing capacity for recovery. Evidence that is even more dramatic comes from patients with isolated cerebral hemispheres, a condition that arises after a so-called split-brain operation. Because this radical operation remarkably preserves function, such patients are invaluable for giving us clues to subtle functional asymmetries and basic capabilities of each cerebral hemisphere.

Research on brain laterality has provided extensive insights into the organization of the human cortex. The surgical disconnection of the cerebral hemispheres (split brain) has produced an extraordinary opportunity to study which cognitive and perceptual processes, are cortical in nature and which, subcortical. Researchers have discovered that visual perceptual and tactile-patterned information; for instance, remain strictly lateralized to one hemisphere following sectioning of the corpus callosum.

When the corpus callosum is fully sectioned, little or no perceptual interaction takes place between the now isolated hemispheres. Surgical cases where callosal section is limited or part of the callosum is spared have enabled investigators to examine specific functions mediated by regions. For instance, when the splenum, the posterior area of the callosum that links the occipital lobe is spared, visual information is transferred normally.

The patterns, colors and linguistic info presented anywhere in either field can be matched with the info presented to the other half of the brain. However, the patients show no evidence of inter-hemispheric transfer of tactile info from objects felt by them. These observations are consistent with other human and animal data showing that major callosum subdivisions are organized into functional zones: posterior ones mediate visual info and the anterior ones transfer auditory and tactile info.

The anterior region of the callosum is involved in higher-order transfer of semantic info i.e. the patient is able to name stimuli presented in the left VF following a resection limited to the posterior callosal region. Upon sectioning of the anterior callosum, this capacity ceased.

By testing each disconnected hemisphere, one can assess the different capacities each might possess. While some claims are exaggerated, there are marked differences between the two halves. The most prominent lateralized function in the human brain is the left hemisphere's capacity for language and speech.

The second primary task domain studied in split-brain subjects is visuo-spatial processing. Here the simple task of arranging some red and white blocks to match those of a given pattern demonstrates poor performance by the left hemisphere while the right one excels.

However, if a picture of the block design pattern is lateralized, each hemisphere can easily find the match from a series of pictures. Moreover, since each hand is sufficiently dexterous, the crucial link must be in the mapping of the sensory message onto the capable motor system.

After the human cerebral hemispheres are disconnected, the verbal IQ of the patient remains intact and although there may be some deficits in free recall capacity, the problem-solving ability of the left hemisphere appears unaffected. However, the right hemisphere seems incapable of higher-order thought processes providing compelling evidence that cortical cell number by itself cannot fully explain human intelligence.

Attentional mechanisms, however, can involve other subcortical systems. Taken together, cortical disconnection produces two independent sensory information-processing systems that call upon a common attentional resource system in carrying out perceptual tasks.

Split-brain studies also reveal a complex mosaic of mental processes that enter into human awareness. As noted above, the two hemispheres do not represent information in an identical manner as evidenced by the fact that each one has developed its own repertoire of specialized capacities.

In the vast majority of individuals, the left hemisphere is clearly dominant for language and speech and seems to possess a uniquely human capacity to interpret behavior or to construct "theories" about the relationship between perceived events and feelings. Right-hemisphere superiority, on the other hand, can be seen in tasks such as facial reawareness and attentional monitoring. Both hemispheres are likely to be involved in the performance of any complex task, but with each contributing in a special manner.

All models of normal language comprehension have to begin with the problem of how words are represented. Understanding spoken language and understanding written language share some processes but there are also some striking differences in how spoken and written inputs are analyzed by the brain.

The input signal in spoken language is very different from that in written ones. Whereas for a reader it is immediately clear that the letters on a page are the physical signals of importance, a listener is confronted by a series of sounds intermingled with others in the environment. He has to identify and distinguish the relevant speech signals from other "noise".

Numerous studies suggest that the superior temporal (ST) cortex is important in mediating sound perception in humans whereas some PET and fMRI studies that have looked at reading of words and pseudowords (sound like words but have no inherent meaning) found middle temporal gyrus activations. This may imply that reading can and often does activate phonological information.

For written input, readers must recognize a visual pattern, which may vary across different writing systems. The symbols used are abstract representations that do not resemble what they represent. The actual identification of such orthographic units may take place in the occipital-temporal regions of the left hemisphere.

The inherent complexity of the language system does hamper the elucidation of its structure and neural mechanisms and so the struggle to clarify them remains one of the great scientific challenges currently facing cognitive neuroscience.

Nonhuman primates also have differences in hemispheric structure or function. In anatomical studies of Old World monkeys and apes, we find asymmetries similar to those of humans. For example, the sylvian fissure shows a greater upward slope in the right hemisphere and there is a similar forward skewedness of the right hemisphere. Whether these anatomical asymmetries are associated with behavioral specializations remains obscure and the case for hemispheric specialization in nonhuman primates is not compelling.

Much unlike humans, the nonhuman primates do not show a predominance of right-handedness. Individual animals may display a preference for one hand or the other but there is no consistent trend for the right hand to be favored over the left either when making manual gestures or when fabricating tools.

Perceptual studies provide more provocative indications of parallel asymmetrical functions in humans and other primates. The rhesus monkeys are superior in making tactile discriminations of shape when using the left hand as in humans. What is more impressive is that split-brain monkeys show hemispheric interactions comparable to those observed in humans on visual perception tasks. In a face reawareness test, the monkeys have a right hemisphere advantage whereas in a line orientation task, the monkeys have a left hemisphere edge.

Complementary studies on patients with focal brain lesions and normal controls tested with lateralized stimuli and even comparative approaches have underlined not only the presence but also the importance of lateralized processes active in awareness and perception.

Recent work, such as the spatial frequency hypothesis has moved brain laterality research toward a more computational account of hemispherical specialization seeking to explicate the mechanisms underlying many lateralized perceptual phenomena. These theoretical advances are taking the field away from the popular interpretations of cognitive style and providing a scientific basis for these robust processes.

The difference in the way the two hemispheres approach the world might also provide some clues about the nature of human consciousness. In the media, split-brain patients have been described as having two brains. The patients themselves, however, claim that they do not feel any different after the surgery than they did before.

They do not have any sense of the dual consciousness implied by the notion of having two brains. How is it that two isolated halves give rise to a single consciousness?

According to Gazzaniga, the so-called left hemisphere 'interpreter' may be the answer. The interpreter is driven to generate explanations or rationales regardless of circumstances. The left hemisphere of split-brain patients does not hesitate to offer explanations for behaviors that are generated by the non-verbal right half even though it does not have access to its information.

In a similar way, the interpreter also does not have a problem with generating spurious explanations for sympathetic nervous system arousal in neurologically intact individuals. Put simply, the left hemisphere interpreter may generate a feeling in all of us that our consciousness is integrated and unified.

Just as a split-brain patient does not realize or find that one side of the brain has gone MIA ("missing in action") and lost all consciousness about the mental processes managed by the right half and vice versa, we don't miss what we no longer have access to.

The emergent conscious state arises out of each side's capacity and probably through neural circuits local to the capacity in question. If they are disconnected or damaged, there is no underlying circuitry from which the emergent property of consciousness would arise.

# Innate Speech

Genetic studies in humans and songbirds are now providing clues about how and when the remarkable human talent for speech arose. The discovery of the so-called FOXP2 "speech gene" created a sensation in the 1990s as it finally provided evidence that the ability to speak is, indeed, written in our DNA.

Interestingly, this gene complex has also been identified in primates, whales, birds and even crocodiles so it is highly likely that all vertebrates have it. Furthermore, the amino acid sequence of this gene is almost identical to that in humans. For instance, only three of the 715 amino acids in the mouse FOXP2 differ from those in the human version. The timing and location of the gene's expression in the brains of other species is also similar.

So what is the FOXP2 gene doing in the brains of these animals, none of which is capable of speech? And, how exactly do these genes regulate a complicated mental process such as vocalizations or speech?

Studies of the FOXP2 gene in people and other animal species, especially songbirds, whose vocal learning resembles that of humans, could help explain why speech evolved in humans but not in any other animal species. Apparently, the FOXP2 gene complex is not solely responsible for the entire human repertoire of vocalizations, but only one of many factors.

Why did scientists label this gene FOXP2?

Actually, "FOX" is an abbreviation for "fork-head box", a characteristic DNA segment present in many genes. Mutations in this gene give the head of fruit-fly embryos a fork-like shape with multiple tines sticking out of the 'head' region. The FOX gene family is so large and branched, that scientists further classify it into subdivisions A through Q. Accordingly, FOXP2 means fork-head box gene family, subgroup P, member number 2. Moreover, FOXP2 has siblings FOXP1, FOXP3 and FOXP4.

Ever since the first published reports about FOXP2, molecular geneticists as well as linguists have been engaged in a heated debate about how precisely the gene affects human speech. Although the gene appears to be crucial for normal development, its specific role remains cloudy. The FOXP2 codes for a specific protein or transcription factor that affects how hundreds or even thousands of other genes are read and translated into their respective gene products.

As a transcription factor, the FOXP2 protein serves as an on-off switch for numerous target genes. Since most genetic material exists in duplicate, a mutation on one chromosome causes the body to produce only half as much of the transcription factor as it should. The resulting shortage somehow causes the spectrum of 'speech' defects found in the affected songbirds or humans.

Mutations to this gene impede the development of certain regions of the brain responsible for not only language processing but also the fine motor control of muscles specific to vocalization. Furthermore, since the FOXP2 gene occurs in a variety of species ranging from reptiles to mammals and birds, it may be mediating other functions besides facilitating speech.

Many researchers are especially interested in studying the avian FOXP2 gene as some of the so-called songbirds learn their songs in a way that is strikingly similar to how human children learn to speak. Initially, baby lark chicks can mimic only the 'cadence' of their future songs. This type of vocalization, a sub-song, resembles the human infant babbling phase.

As the chick matures and hears an example of what the song should really sound like from an adult lark or a parent, it adapts its vocal repertoire to reflect its new mastery. Just like human children, songbirds are dependent on what they hear to develop 'normal' vocalization (local accents).

The similarities between learned avian song and human speech run even deeper. Both humans and songbirds have developed neuronal structures that specialize in the perception and production of sounds. Compared with humans, whose brains use parallel processing to comprehend speech, songbirds have a rather more modularly constructed brain in which various centers assume specific roles.

In the avian brain, auditory stimuli reach the high vocal center, which controls the muscular movements of the vocal organ via the motor center. (Any damage to this region prevents singing.) Another important data path in the birdbrain loops from the high vocal center via area X, (a song-learning center in the basal ganglia) to the thalamus and back to the cortex. This so-called cortico-basal ganglia loop also exists in several mammals including humans and mediates learning.

Massive amounts of FOXP2 proteins are produced in the basal ganglia, which is also where the structural and functional anomalies occur in patients with FOXP2-

related speech defects (incoherent mumbling, stumbling over grammar, inability to describe events in a correct chronological order and other speech impediments).

The logical implication of some of these findings suggests that the evolution of language is not a unique feature of the human lineage. Many species share the structure and molecular make-up of the brain that was already in place as our hominid ancestors began to speak. Only as existing genes and neuronal systems continued to develop was the path cleared for the uniquely human capacity for speech.

# Birth of Sentience

Each species has evolved its own modus operandi based on the complexity of their nervous system. This simple fact implies that even though all the various species inhabit the same world, they tend to interact with that environment differently. Bees 'see' in UV light, snakes sense infra red radiation, birds navigate using cues from the Earth's magnetic field, fish possess sense organs sensitive to vibrations in the water while dogs have hyperacute smelling and hearing abilities.

So when in evolution did consciousness first appear?

Well, it all depends on how you define consciousness!

Simple awareness of the surroundings and environment arose quite early in animal evolution. Internal homeostatic mechanisms dependent on numerous sense organs were in place by the bacterial or amoeba stage of evolution.

This was followed by 'avoidance and predatory mechanisms' where the organism either fled the scene or attacked, depending on how its sensory organs 'assessed' the situation. There was no true 'understanding' or 'insight' here. In Humphrey's words, they "went about their lives, deeply ignorant of an inner explanation of their own behaviour."

Next came 'intelligence', where the animal is now capable of not only assessing, but is also able to mount a graded response via a true analysis of the situation. Analogous to the action of a dimmer switch, intelligence comes gradually in increments and degrees, with increasing complexity of the underlying neural assembly.

In the so-called higher mammals and primates, innate intelligence evolved into a true understanding of consequences and a rudimentary concept of self. As any pet owner will attest, kittens will rush up to a mirror as if looking for its playmate. Many birds continue to carry-on and peck at their image as if confronted by a hostile rival. Even monkeys clearly fail to 'understand' that the reflection is their own.

The great apes, on the other hand, gorillas, chimps, orangs and humans, quickly realize this and recognizing themselves in the mirror, begin to 'inspect' various hidden attributes of their body. Interestingly enough, human babies, below two years of age also fail to recognize themselves in a mirror. Teenagers, on the other hand, cannot get enough of their reflection and are always seeking out mirrored surfaces in shopping malls.

According to Dennett, a cognitive scientist at Tufts University, sometime during the second year or so, babies begin to follow another person's direction of gaze rather than their pointing finger. Numerous experiments demonstrate that between the ages of three and five, humans develop a so-called 'theory of mind' i.e. begin to understand other people's behavior as 'intentional'.

They attribute mental states to others or to themselves that govern hope, fears, wants and needs. This aspect of consciousness is absent in most other advanced mammals and is seen as a very powerful 'tool' for understanding, manipulating and predicting the world around us. It also heralds the birth of true self-consciousness.

It is plausible that some animals notably mammals possess some but not necessarily all features of the consciousness spectrum via their sensorium tailor-made to navigate efficiently in their ecological niche. Many people that have pets and work closely with animals intuitively assume that they have their own flavor of subjective states. To believe otherwise is not only presumptuous, but flies in the face of experimental evidence for the continuity of behavioral patterns between animals and humans.

This is particularly true for monkeys and apes, whose behavior, development and brain structure are remarkably similar to those of humans. In fact, one of the best ways to study stimulus awareness today relies on correlating neuronal responses of trained monkeys to their behavior.

Obviously, humans do differ fundamentally from all organisms in their ability to speak. True language enables *Homo sapiens* to represent and disseminate arbitrarily complex concepts. Unfortunately, the primacy of language in the human social milieu has given rise to a belief among philosophers, linguists and other theoreticians that consciousness is impossible without a true language and that therefore, only humans can feel and introspect.

While this may be partially true, most evidence from split-brain patients, autistic children, evolutionary studies and diverse animal behavior observations does suggest that most of the "higher" mammals do cognate and experience various degrees of pain and emotions. At present, it is unknown to what extent conscious perception is common to all animals. It is possible that consciousness correlates to some extent with the complexity of its nervous system.

Squids, bees, fruit flies and even roundworms are all capable of fairly sophisticated behaviors. Perhaps they too possess some level of awareness, feel pain, experience pleasure and plan foraging routines. It would be contrary to evolutionary continuity to believe that consciousness is unique to humans. As we will see throughout this book, the human mind does share some basic properties with animal minds.

Stereo-taxic studies performed on animal subjects with micro-electrodes, reveal segregated cortical neighborhoods whose neurons are specialized to carry out different jobs. For instance, neurons in one occipital-temporal region are particularly sensitive to the color or hue of stimuli. Neurons in the MT region detect movement in the environment whereas neurons of the posterior parietal cortex program fine eye movements to track them. Clinical observations of patients suffering from mild neurological deficits reinforce the view that particular regions of the cerebral cortex subserve specific functions.

Cortical areas in the back of the brain are organized in a loosely hierarchical manner with at least a dozen levels, each one subordinate to the one above it. When a group of neurons within one of these regions receives a strong driving input from lower in the hierarchy, the neurons send their output to another area or group located higher in the hierarchy. Feedback loops abound and numerous 'short-cuts' are present within this structured assembly producing a flexible "coalition", if you will, rather than a rigid hierarchy.

The incoming sensory information is also not usually enough to lead to an unambiguous interpretation. In such cases, then, the cortical networks fill in. They make their best guess, given the incomplete information. This filling-in occurs throughout the brain and guides much of animal or for that matter, human behavior. Any visual scene gives rise to widespread activity across the brain.

Coalitions of neurons, coding for different objects in the world out there, compete with each other i.e. one coalition strives to suppress or inhibit others, vying for a subjective response. This is particularly true in the higher centers of the cortex. Paying attention to an event or object biases this competition in its favor leading to the emergence of a winner. The activity associated with the winning coalition corresponds to the conscious state.

Usually only a single coalition survives but in many cases when the neuronal representations do not overlap, several may coexist alongside for a while. The losing coalitions do not, however, lie down and die. Like inveterate politicians, they continue to remain active and influence the cortical hierarchy, jockeying to win the next election.

The essential difference between the twin processes of awareness and consciousness, as I see it, using the terminology of today, is the difference between a tangible computer system and virtual cyberspace. The brain along with its neuronal elements symbolizes a modern computer and the intricate "web" constructed by it to do it's computation in, is what we call consciousness.

Just as the computer requires cyberspace to function within, the brain constructs this elaborate virtual workspace, to perform its duty.

Does that mean that consciousness is a mere epiphenomenon brought about by a functioning brain or complex neural network?

I'm not sure.

Is it some kind of metaphoric space, stage or Cartesian Theater within which awareness occurs?

Possibly.

So where is this place or space in the real Einsteinian SpaceTime continuum?

It is hard to explain or pin down, just like cyberspace. Nobody really knows what or where it is but everyone works in it whenever any computer terminal is used. It is definitely not real physical 3D space but virtual or made-up space, a figment of our imagination!

However, this is where all the "mental stuff" takes place: trances, meditation, OBEs, NDEs, paranormal phenomena, ESP etc., etc.

# Human Consciousness

There is much disagreement over how and when consciousness first appeared on Earth. Some believe that its appearance was gradual, increasing like a dimmer switch, if you will, with each increment in brain size. Others, like the pan-psychists, believe the entire universe is conscious, although there are degrees of it ranging from the simple (rocks, bacteria and viruses) to increasingly complex (animals and primates).

In this view, consciousness itself came long before biological evolution began and gradually evolved into the higher forms. Some believe that life and consciousness are inseparable and emerged together four billion years ago. Others think that consciousness requires a complex nervous system and that it must have evolved along with the increasing complexity of brains.

Finally, there are those who feel that consciousness dates from the increasingly social milieu of our early ancestors, circa two million years ago, which induced them to build novel skills of understanding, predicting and manipulating the behaviors of others.

An interesting implication of this hypothesis is that only intelligent and highly social creatures are conscious i.e. most creatures prior to such creatures through much of the Earth's history were not conscious at all. Their 'brains' processed information from the sense organs without 'thinking' and their bodies acted to avoid fear and hunger, without their 'minds' being conscious of any accompanying emotions.

How does the brain convert sensory signals into a coherent perception of the world out there?

Considering the complexity of the nervous system, it is amazing how easily we take in all the familiar sights and sounds of the neighborhood to form a coherent precept. In the next few pages, we will explore the big picture, if you will, of the cognitive neuroscience of perception.

We have gotten some clues about how consciousness emerges from the study of so-called "split-brain" patients, whose corpus callosum is severed to provide relief from severe intractable epilepsy. The obvious fear was what the side effects of this radical surgery would be.

Would it result in a "split" personality in that patient since there were now two chimp-size brains sitting inside the skull?

Or, would the two halves of the brain vie for 'dominancy' to determine which one is in charge in particular everyday situations?

Although what we commonly call "mind" is far from being a synonym for consciousness, it tends to influence the degree of consciousness we experience during the day. For example, you will notice a quickening in your step when getting home from a boring day at work, to a roaring block party thrown by close friends.

Awareness has many levels and it seems that the many separate stages follow a sort of neurological "hierarchy", if you will, with the more complex ones receiving input only after it has undergone processing by the more elementary systems. Every synapse along the lengthy pathway adds an element of organization and seems to contribute to the overall sensory experience.

Intuitively, all the neuroscientists have converged on the brain as the only anatomical entity that could possibly mediate and sustain such awareness or primary consciousness.

In their view, the term consciousness includes two distinct concepts:
(1)     States of Consciousness (SOC), and
(2)     Conscious experience i.e. thoughts and emotions,
during the SOC.

In my opinion, they are looking for it in all the 'wrong' places. The human CNS, like the Empire State building, is the culmination of a long lineage of biological architecture and mental refinement. We need to keep in mind that the CNS evolved. It was not created en masse by some supernatural Agency or Creative Force. In order to gain a basic understanding of 'dwellings' one needs to study an animal 'shelter' first.

We need to go 'down' the phylo-genetic scale of earthly organisms and 'decide' when true consciousness made its debut; whether it is at the cellular cytoskeleton level in Amoebas, or further up when the notochord (primitive central neural axis) first appeared, in Chordates.

In a unicellular organism, like the bacterium or even the amoeba, all these neurological states are induced within the confines of its cell wall (triggered by various chemical 'messengers'), analogous to the same 'one room shelter' serving as a kitchen, living room or bedroom depending on the situational demand.

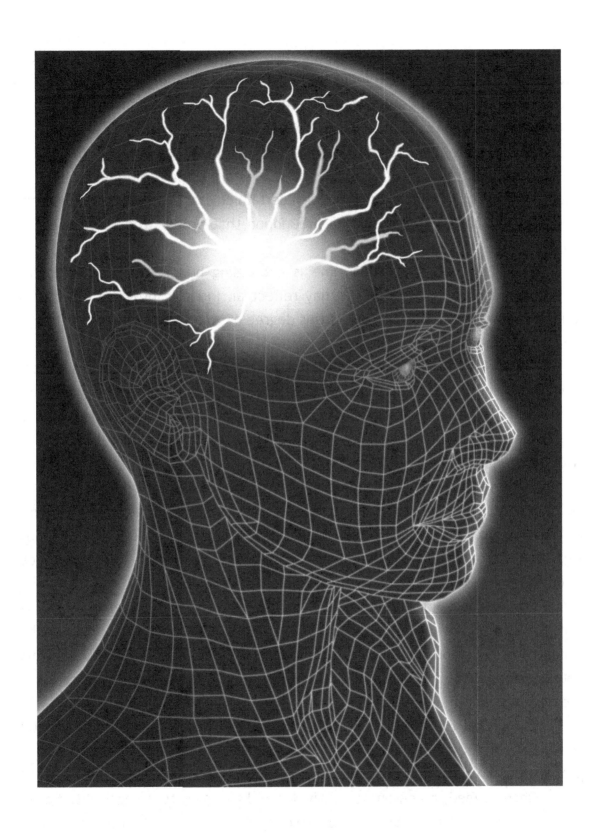

Cells are in a way more complex than an embryo. The reason is that the network of interactions between proteins and DNA within any individual cell has many more components and is consequently much more complicated than the interactions between individual cells of a developing embryo.

For example, when a cell receives a signal at its membrane from another cell, it precipitates an entire cascade of protein-protein interaction, which may lead to the opening of an ion channel or a certain cluster of genes being switched on/off or the abrupt extension of a pseudopod that propels the amoeba out of danger.

Any given cell at a given time is expressing hundreds if not thousands of different genes much of which may reflect an intrinsic program of activity, independent of external signals. It is this degree of complexity, which determines how cells respond to various stimuli.

This state can reflect the cell's development history – cells have good memories – and so different cells can respond to the same signal in very different ways. Numerous researchers have noticed many examples of the same signals being used repeatedly by different cells at different stages of embryonic development with different biological outcomes.

The evolution of multicellular forms of life is the result of changes in development, which in turn is due to changes in genes that control cell behavior in the embryo. Nothing in Biology makes sense unless viewed in the light of evolution. Certainly, it would be quite a challenge to make sense of many aspects of development without an evolutionary perspective.

For example, despite different modes of very early development, all vertebrate embryos develop through a rather similar phylotypic stage after which their development diverges again. This shared stage, which occurs after neurulation and the formation of the somites, is a fundamental characteristic in the development of all vertebrates.

The neural tube generates a large number of different neuronal and glial cell types. In both the brain and spinal cord, all neurons and glia arise from a proliferative layer of epithelial cells lining the lumen of the neural tube.

Once formed, a neuron never divides and migrates to its designated area. Growth cones at the tip of the extending axon guide it to its destination. Filopodial activity at the growth cone is influenced by environmental factors such as contact with the substratum, other cells or subtle gradients in cell-surface molecules (chemotaxis).

# Primary "Consciousness" or Heightened Awareness

Let us first identify the stage in human development or in the evolution of species, when we can expect a brain to be capable of expressing primary consciousness. The most plausible scenario is that this so-called consciousness begins in the womb. Since at conception, the fertilized egg being merely a large cell with no CNS or brain, possesses no wherewithal for sentience.

During embryogenesis, even as generic stem cells begin to assume neuronal duties, there is no set threshold after which 'consciousness' comes into play. All number of key events in the developing brain unfold during the first eight weeks of the growing fetus without bestowing any obvious, outward "sign" of increasing sentience.

How about when the brain cortex develops, since in mammals, this region of the brain has been implicated in emerging consciousness?

Unfortunately, there are many stages to the laying down, if you will, of the neocortical neuronal "sheet" upon the midbrain "stalk", spanning many months (from six weeks after conception to about thirty weeks when the first sign of an EEG pattern becomes discernible.

The same riddle of isolating the time or place when consciousness "happens" applies to all non-human creatures. Most people would agree that it is highly unlikely that although human are considered conscious, chimps, who are less than 2 percent different in their genetic information, are insentient zombies.

And moving still further away from our own species, most pet owners would take exception to their companions being considered anything but warm, intelligent and compassionate creatures. In other words, if an animal can have discernibly different global states of something, then surely that "something" cannot be anything other than some 'crude' or primary type of consciousness.

Hence, if the human fetus is considered "conscious" by virtue of its newly minted CNS and brain structures, then this consciousness, analogous to all other non-human primates, will, by definition, have to be somehow different in degree from what we as adults or even as children, experience in our daily lives.

I agree with Dr. Greenfield's suggestion that the process of consciousness should be viewed as an on/off or there/not there entity but rather a continuum: not as a sudden blinding light but rather as a dimmer switch, able to ramp-up the spotlight of awareness in gradual increments.

This switch would vary from almost imperceptible light (read basic awareness as in parameciums) to the floodlit brilliance of adult humans. Quite simply, consciousness would grow in degree as brains do. This will enable it to expand not just during the growth phases of childhood and infancy but throughout its lifetime.

Deepening consciousness would then imply an awareness of a greater significance in the world around you, intimately determined by your life experiences, ingrained biases and prejudices that have been instilled into your psyche on your journey into adulthood.

# Intuition & 'Gut' feelngs

As studies over the past decade have confirmed, much of our information processing occurs below the radar, if you will, of our awareness. However, the extent to which these seemingly nonconscious, automatic mental processes pervade all aspects of our conscious social existence is very difficult for the ordinary person to comprehend.

We naturally assume that our intentions and choices govern our lives but that is only partly true. We tend to overrate how much control we really have over mundane existence. In reality, most of us muddle through life mostly on autopilot, as it were.

By studying the forces that shape our intuitions, the brain scientists have begun to piece together how our 'hidden' mind mediates not only our insights and creativity but also our implicit prejudices, biases and irrational fears.

So, what is intuition?

Psychologist Daniel Kahneman in his 2002 Nobel Prize lecture noted that our 'consciousness' appears to follow a dual track Jekyll & Hyde personality. Our behind-the-scenes intuitive mind is fast, automatic, effortless, associative, implicit (not available to introspection) and often emotionally charged. Our familiar, conscious mind, on the other hand, is deliberate, sequential and rational requiring an effort to engage.

According to Kahneman et al., humans have evolved mental 'shortcuts' (heuristics) that enable efficient and 'snap' judgments on evaluating scant or 'gist' data. Such heuristics are like perceptual cues that usually work well but can occasionally trigger illusions, delusions or misperceptions. For example, when a front-desk receptionist is, surreptitiously replaced by another individual lurking beneath the desk, most people fail to notice the switch.

A second influence on our intuitions comes from learned associations, which spontaneously surface as 'feelings' that sway our judgments. Our life experiences provide us with a vast repertoire of behavioral and adaptation skills that emerge during decision-making or responding to a given situation. When speaking, for example, we communicate intended meaning with instantly organized string of words that somehow effortlessly flow out of our mouth accompanied with appropriate hand gestures and body 'language'.

Most of us commonly think of our vision as one integrated system that 'controls' our visually-guided actions. Actually, as described earlier, the process of vision consists of at two systems, each with its own hub in the cerebral neocortex. In their widely publicized studies from the early 1990s, social psychologist Nalini Ambady, then at Harvard and psychologist Robert Rosenthal of UCal, Riverside demonstrated that we often form positive or negative impressions of people we encounter on a daily basis in a mere "blink" or "thin slice" of time.

After subjects observed three two-second video clips of professors teaching, their teacher-ratings accurately predicted the actual end-of-term ratings by the professor's own students. In order to get a good sense of someone's energy and warmth, the researchers found, a mere six seconds will often suffice lending credence to the old adage of 'first impressions being lasting impressions' or judging the book by its cover.

Thanks to visual pathways that run from the eye to the brain's rapid-response emotional-control centers – bypassing the 'thinking' part of the brain – we often 'feel' before we analyze. If experience informs our intuition, then as we learn to associate cues with particular feelings, many judgments should become automatic. Here some studies have shown that women do have an edge in spotting lies or seeing through a display of 'fake' emotions.

Yet psychological science is replete with examples of smart people making predictable and sometimes costly intuitive errors. These occur when our experience has exposed us to an atypical sample or when a 'quick-n-dirty' heuristic leads us astray. For example, various experiments comprising of briefly flashed words or faces that 'prime' or automatically activate stereotypes for some racial, gender or age group is often shown to 'bias' the participant's behavior or response.

Even the most seemingly tolerant, egalitarian 'whites' are shown to take longer to identify pleasant words (such as peace and paradise) as good when associated with black rather than white faces. Moreover, the more strongly people exhibit such implicit prejudice, the readier they are to perceive anger or 'shiftiness' in black faces.

Finally, why do we fret about remote possibilities while ignoring higher probabilities? In other words, why does a chain-smoker worry about riding an elevator more than his deadly habit? Why do we fear violent crime more than obesity or clogged arteries? Why do women fear breast cancer over heart disease?

Such questions reveal that intuition is powerful and often wise but also sometimes perilous, especially so when we 'overfeel' and 'underthink'. What today's cognitive science does is enhance our appreciation for intuition but also reminds us to check it against reality. Smart, critical thinking often begins as we listen to the creative whispers of our vast unseen mind and builds as we evaluate evidence, test conclusions and plan for the future, rationally.

Of course, the obvious cure for the tragic shortcomings of human intuition in a high-tech world of today is better or should I say, more relevant education. Due to various time constraints in our busy lives, the question is not whether an educated person should learn a foreign language or know the classics but attempt to step beyond the individual 'comfort zone'.

This can be achieved by setting priorities for the revamping of our outmoded and archaic educational policies and curricula so that our high school students will be equipped with a set of more appropriate cognitive tools. These tools would then, be utilized to thrive in a world whose complexities are constantly challenging our intuitions and where subtle every-day tradeoffs cannot responsibly be avoided.

# Unity of Consciousness

Since so many of neurological studies implicate the immensely diversified brain as being the main organ system mediating our states of consciousness, why do we feel as though there is just one 'mind' experiencing a unified world out there? In other words, there is one constant and undifferentiated self, 'me', perceiving the world through one continuous 'stream' of experiences occurring around 'me'.

Our experiencing self, (awareness?) seems to be at the center of everything we are aware of at a given time and to be continuous from one moment to the next. It thus seems to have both unity and continuity: an inner agent that carries out actions, makes decisions, bestows a unique personality to the Self and becomes the sole determinant of your lifestyle.

However, is this perception... false?

Do we really have an underlying, unitary Self, separate from their physical brain (Dualism)?

Or... is it, as the Buddha claimed, that there is no such thing as the Self, but just a name or label given to something that does not exist (Monism)?

Or... we could simply reject the whole idea that consciousness is a unified entity and call it as being just a big delusion, induced by the machinations of *Maya Jahl* ( Sanskrit for Web of Delusion)?

What do some of the modern "split-brain" patients reveal when the 200 million axons constituting the corpus callosum (inter-hemispheric linkage) are surreptitiously severed?

Surprisingly, most seem to be completely unaware of any changes in their mental processes even though they will do subtle things to compensate for their loss of brain connectivity. They may move their heads to feed visual information to both hemispheres or talk out loud for the same purpose or make symbolic hand movements that afford clues for slight deficits in perception.

After the right and left half of the brain are "disconnected", the verbal IQ of a patient remains intact and so does his problem-solving capacity. There may be some deficits in free-recall capacity and in other performance measures, but isolating essentially half of the cortex from the dominant left half causes no major change in cognitive functions.

(**Author's Note**: Most of the subcortical pathways do remain intact. Both hemispheres are still connected to the brain stem, so both sides receive much of the same sensory and proprioceptive information which automatically codes the body's position in 3D space. Both hemispheres can initiate eye movements and the brain stem supports similar arousal levels so both sides sleep and wake up at the same time.

There also appears to be only one integrated spatial-attention system, which continues to be unifocal as attention cannot be distributed to two disparate locations. Emotional stimuli presented to one hemisphere will still affect the 'judgment' of the other.

Although both hemispheres can generate spontaneous facial expressions, only the dominant left one can do so voluntarily. Also, since 'half' the optic nerve crosses from one side of the brain to the other at the optic chiasm, the information from regions of the retina that usually attend to the right visual field will not cross over.

For example, if a patient sees something in isolation in the left visual field, only the non-verbal right side of the brain has access to that visual information. And since it cannot share this info with the verbal left hemisphere, the patient, when asked to elaborate, tends to "interpret" or make up a plausible story.)

Popper and Eccles argue that the mind plays an active role in selecting, reading out and integrating neural activity, molding it into a unified whole according to its desire or interest.

But... how?

No explanation to this so-called 'binding problem' is given.

Most of the other 'theories' attempting to explain the dynamic binding taking place in real time, in the complex process of consciousness, have concentrated on the human visual system.

The best known is the one proposed by Crick and Koch, where stereotaxic studies of the cat's visual cortex revealed oscillations in the 35-70 hertz range, all firing in 'synchrony'. They suggested that "this synchronized firing, on or near the beat of a 'gamma oscillation' (35-70 hertz), might be the neural correlate of visual awareness."

Subsequent research in other animal species, including humans, has demonstrated that neuronal synchrony may be related to perceptual integration, the construction of coherent representations, attentional selection and awareness.

Despite all the exciting advances in systems neuroscience and with the insight that many of our cognitive capacities, so heavily a part of our conscious experience, appear to be built-in domain-specific operations, we think we are a unified conscious agent with a past, a present and a future.

These domain-specific or modular systems are fully capable of producing behaviors, mood changes and cognitive activity. So with all this independent activity, what allows for this sense of conscious unity that we can "self"?

It turns out that we, as humans, have a specialized system to carry out this interpretive synthesis and it is located in the left hemisphere. Known as the "interpreter" by Gazzaniga and LeDoux, its discoverers, this system seeks explanations for how contiguous events relate to one another. It operates on other adaptations built into our brains. The adaptations are most likely cortically based but they work largely outside of conscious awareness, as do most of our mental activities.

Gazzaniga and associates wanted to know whether the emotional response to stimuli presented to one-half of a split-brain patient would affect the other half. They found that the emotional valence of the stimulus clearly crossed over from the right to the left hemisphere. The left remained unaware of the content that produced the emotional change, but it interpreted and experienced the emotion.

For example, if the command "walk" is flashed to the right half, the patient typically stands up and heads for the door. When asked where he is going, the patient shrugs and says, (using the left half to analyze and speak) "Oh, to the vending machine for a soda."

The brain's modular organization has now been well established. The functioning modules do have some kind of physical instantiation, but the brain scientists cannot yet specify the nature of the neural networks. However, what is clear is that these networks operate mainly outside the realm of awareness and announce their computational products to executive systems that mediate behavior or cognitive states.

Catching up with all of this parallel and constant activity appears to be the responsibility of the left hemisphere's interpreter module. The interpreter is a system of primary importance to the human brain: It allows for the formation of beliefs, which in turn are mental constructs that free us from simply responding to stimulus-response aspects of daily life.

With this view, it becomes evident that what we mean by consciousness is how we feel about our specialized capacities. Our consciousness is essentially an inferential system that collates all sorts of mental activity and assigns feelings to them. It reflects the affective component of specialized systems that have evolved through the intricate process of natural selection to enable human cognitive processes and become personalized "qualia".

# The search for 'Qualia'

Much of our understanding of the animal CNS (Central Nervous System) leads us to the classic picture in which neurons, consisting of a cell body housing a nucleus, extend out into a long un-branched axon at one end and a profusely branched dendrite at the other. The dendrites receive stimuli and conduct impulses generated by those stimuli toward the cell body. There is some evidence of 'dentritic processing' taking place. The cell body then sends out a 'graded response' out through the axon to the specific tissue for execution of the proper action.

Communication from neuron to neuron occurs at the synapse, where a terminal branch of the axon of one neuron makes contact with the cell body, dendrites or axon of another neuron. This communication can occur through electrical or chemical synapses. The electrical synapses consist of low-resistance connections between the membranes of two neurons where a change in potential in a pre-synaptic neuron transmits to a post-synaptic one with little attenuation. Electrical synapses are ubiquitous in the invertebrate and lower-vertebrate nervous system and at some sites in the mammalian nervous system.

The chemical synapses are the predominant type of inter-neuronal communication in the mammalian brain. Action potentials in the pre-synaptic neuron cause the release of neurotransmitters or neuro-modulators from synaptic vesicles. These 'chemical messengers' traverse the synaptic clefts and mediate excitatory or inhibitory responses on the post-synaptic membranes.

As is evident from the above description, there is a distinct 'qualitative' difference between the 'simple' electrical synapses found in the 'lower' life forms and the more complex synaptic terminals of 'higher' ones. The neurons connect to each other locally to form a dense network in portions of the brain called gray matter; they communicate over longer distances via fiber tracts called white matter.

The cortex itself is a six-layered structure with various connection patterns in each layer. There is an enormous amount of variability and uniqueness in each individual brain, which provides the organism with an organizing principle for the proper assessment of the environment it, is in. This perceptual categorization takes place by means of so-called global mappings that connect various modal maps stored in its memory.

The cortex subdivides into regions that mediate different sensory modalities, such as hearing, touch and sight. There are other cortical regions dedicated to motor functions that ultimately drive our muscles. Beyond the sensorimotor portions concerned with gathering information about the outside environment and modifying the internal one to meet those demands, there are regions of the brain such as the frontal, parietal and temporal 'lobes' that 'talk' to each other and maintain homeostasis in the physical body.

Many neurologists now believe that the modulatory neurons confer some permanent structural change on the brain cells that they contact. This change may constitute the same fundamental mechanism of learning and memory at all levels of the animal kingdom. Memory is a central component of the brain mechanisms that mediate an SOC.

It is commonly assumed that memory involves the inscription and subsequent storage in so-called 'association areas'. However, what is stored?

Is it an encoded message in some cryptic neuro-molecule that is yet to be discovered?

Or... Is it simply a complex jigsaw puzzle, hastily reconstructed from the coordinated input of disparate areas of the human cortex?

These queries point to the widespread assumption among the laity that what is stored in your 'head' is a long 'videotape', which has meticulously recorded every event of their lives.

The Edelman-Tononi team takes a different view. They shy away from depicting memory as the inscriptions on the tabula rasa. Instead, they describe memory as the ability of a dynamic system to repeat or suppress a mental or physical act if deemed inappropriate. They illustrate this novel and rather intriguing view of memory as similar to the melting and refreezing of glacial ice. Just as water undergoes multiple transition phases as the ambient temperature hovers around its freezing point, our memory has properties that allow perception to alter recall and recall to alter perception.

It has no fixed capacity limit since it actually generates information by 'constructing' it. It is robust, dynamic, associative and adaptive. It is one of the essential components of this complex biological process of Consciousness. By virtue of being 'creative' rather then merely replicative, it ultimately gives rise to 'Qualia', the so-called higher-order subjective responses to things we perceive. These when bound together into a unitary concept constitutes an SOC.

Primary consciousness arises as a result of circuitous interactions between brain areas mediating value-category memory stores and those 'association areas' of the cortex which derive perceptual precepts based on a weighted average of the incoming sensorimotor data. A consequence of such interactions is the 'construction' of a perceptual 'scene' out there and the ability to respond appropriately to it. This discriminatory capability within a unitary scene is often based on Qualia, the custom-made higher-order and privileged properties of consciousness in any given organism.

In humans, just as the separate snapshots in a film reel are translated into a smoothly flowing movie, the circuitous integration of neural input via numerous cortical areas, give rise to a unitary precept of Qualia. It cannot be divided into separate parts as it is experienced. Instead, the conscious 'scene' is unified and coherent. It is not possible, willfully or with even a high degree of attention to limit awareness to just one component of a scene to the exclusion of all others. Yet, myriad conscious states and scenes can be experienced sequentially with no subjective evidence of discontinuity or disruption. Consciousness itself is an internally constructed phenomenon.

In other words, although perceptual input is initially important, the integration of various areas and modalities in the cortex allows the individual to go beyond the information given. This going beyond, with the appearance of 'filling-in' and gestalt properties involving shifting dominances between cortical maps can be seen in various visual, auditory or somatosensory illusions to which all of us are prone. The same is true of the sense of time, of succession and of duration. The neural integration tends to combine concepts and precepts with its experiential memory stores to construct a coherent, self-consistent picture.

This 'phenomenal transform' of the neural input to the subjective concept of Qualia is one of the major origins of the subjective sense and in humans the notion of self-consciousness. Christof Koch in his book on the Quest for Consciousness theorizes that most people would reflexively assign consciousness to the top of the processing pyramid, which starts with the eyes, ears, nose and other sensors and ends with the 'conscious me' endpoint of all perception and memory.

In his opinion, this view is false; a mere 'cherished chimera'. Instead, he speculates that the neural correlates of consciousness (NCC), may reside in a so-called cognitive architecture straddling a representation of physical objects and events and an inner, hidden world of thoughts and concepts. He invokes Ray Jackendoff's analysis, which assumes a tripartite division of the mind-body into the physical brain, the computational mind and the phenomenological mind.

In his book Consciousness and the Computational Mind, Jackendoff defends a so-called intermediate-level theory of consciousness. He argues that even though common sense suggests that awareness and thought are inseparable and that introspection reveals the contents of the mind, both these beliefs are untrue. According to him, thinking i.e. the manipulation of concepts, sensory data or more abstract patterns, is largely not conscious.

What is conscious about thoughts, are images, tones, silent speech and to a lesser degree, other bodily feelings associated with intermediate-level sensory representations. Neither the process of thought nor its content is knowable by consciousness. In other words, you are not directly conscious of your inner world, although you have the persistent illusion that you are!

Koch elaborates: "you are only conscious then of representations of external objects (including your own body) or internal events by proxy. You are not directly conscious of something in the world, say a chair, but only of its visual and tactile representation in the cortex. The chair is out there; your only direct knowledge of it is derived from explicit but intermediate representations of your senses in your brain that leave out many fine details...

Likewise, recalled or imagined things are, mapped onto visual, olfactory, gustatory, vestibular, tactile and proprioceptive representations. A subset of these, the NCC, is experienced as Qualia. These involve one or more dominant coalitions stretching between cortical regions in the back and more frontal ones."

Whether the Qualia associated with these feelings, more diffuse and less detailed than sensory percepts, exist in their own right or are mixtures or modifications of various bodily sensations, is unclear.

Some researchers, Dennett for example, assert that sentient experiences or qualia are a cognitive illusion (there's that word again). Once we have isolated the computational and neurological correlates of access-consciousness, there is nothing left to explain? He feels that it is irrational to insist that qualia remain unexplained even after all its manifestations have been accounted for.

To him this is like insisting that wetness remains unexplained after all the manifestations of the water molecule have been described in detail. Well, as a sentient human, the unexplained quality of wetness does leave behind a feeling of dissatisfaction for me. It implies that just because we do not yet have a scientific explanation of cognitive perception, it does not exist.

In my opinion, Dennett is trying to banish this phenomenon from discourse by invoking the doctrine of logical positivism i.e. if a statement cannot be scientifically verified, it is essentially meaningless. He is really not explaining qualia as much as explaining it away.

The presence of disparate altered states that are not merely transient, as in drug-induced, but which profoundly affect changes in the inductee's life occur as so-called 'exceptional human experiences' (EHEs). These encompass psychic visions, lucid dreams, 'out-of-body experiences' (OBEs), 'near-death experiences' (NDEs) or various other 'mystical'/spiritual experiences.

While undergoing such episodes, the ordinary sense of a conscious Self, seems to undergo a dramatic disruption. The person emerges quite transformed; often with a different outlook on life, a changed belief system and a reduced fear of death. Some of these experiences prompt paranormal or supernatural claims. Others may be taken as evidence for the existence of an omnipotent Deity, souls, spirits or life after death. What they surely do is raise numerous questions in the scientific world.

Are any of these claims valid? Is there a common thread running through these mystical experiences or are they just a loose collection of unrelated oddities?

Certain psychoanalytic theories describe the OBE as a dramatization of the fear of death, which Jung saw as part of the process of individuation. They find that OBE's tend to predominate when there is a mismatch between the sensory input to the brain and the individual's body image.

When a weak electric current was passed via a micro-electrode into the right angular gyrus of the temporal lobe, the patient reported a 'sinking' or 'floating' sensation that increased with current intensity.

She also reported on various body image distortions during the stimulation and felt like she was detached from her physical body and observing everything being done to her from the ceiling. Interestingly enough, epileptic patients with lesions of the temporal lobe report more such OBE experiences.

Author's Note: Recently, scientists at Sweden's Karolinska Institute created an OBE in the lab by using carefully-placed video cameras. Volunteers sat with their back to two video cameras and wore a special headset that displayed the camera's output, mimicking their normal vision.

When the researchers touched the subject's chest and simultaneously performed a similar motion just below the camera's field of view, the volunteers suddenly felt like they were watching their own bodies from behind.

It is hoped that by further studying this kind of "sensory illusions", the researchers will gain insights into the nature of consciousness and the so-called "first-person" experience.

Near-death experiences, on the other hand, are commonly taken to be proof-positive of the existence of "a higher spiritual world", or a soul that can survive death by leaving the body. An alternative, more naturalistic approach to understanding this process of NDE, known as the "dying brain hypothesis", was introduced in 1993 by Susan Blackmore of Oxford University. She mentions that severe stress, extreme fear and high degrees of cerebral anoxia all cause cortical disinhibition and spasmodic brain activity.

Bright images of lights and tunnels are frequently caused by disinhibition in the visual cortex and the so-called 'life reviews' reported by accident victims can be induced by stimulation of the temporal lobe. The positive emotions and apparent lack of pain were attributed to the action of endorphins, encephalins or endogenous opiates released under extreme stress.

These visions of other worlds and spiritual beings may be real glimpses into the Void, except for the intriguing cultural bias that seems to exude from the description of such episodes. For example, Hindus relate darshans of their own deities decked out in fine regalia, the Parsees describe visions of Ahura Mazda, and Christians greet Jesus in a restored Garden of Eden, or by the pearly gates, with evidence of bloody stigmata in plain sight and the Muslims describe lucid visions of vestal virgins cavorting in a paradisiacal oasis.

Mystical experiences are rather diverse and almost impossible to define in succinct terms. In William James' opinion, the bottom line is, "the whole concern of both morality and religion is the manner of our acceptance of the universe". He proposed four hallmarks of a mystical experience. Ineffability (the inability to describe it in words or successfully impart its emotional content to others), transience (tendency to not last too long), passivity (the mystic feels that a 'higher force' is in total control) and possessing a so-called 'noetic' quality, a term that James coined to describe a superior state of knowledge, insight or illumination.

Deep mystical experiences seem to transcend narrow parochial religious views. In the Essentials of Mysticism, Evelyn Underhill sought to 'disentangle the facts from ancient formulae used to express them'. She defined the essence of mysticism as the clear conviction of a living god in unity with the personal self. Underhill specifically rejected paranormal claims as missing the point. Even Buddhist initiates are taught to ignore visions, miracles and faith healing. "They are no better than dreams which vanish forever on awakening".

Dismissing such reports/experiences as fabrications or wish fulfillment, even though easy, is unreasonable. The similarities observed across the ages and diverse cultures along with the inherent reliability of ones that underwent meticulous monitoring in various labs around the world seem to suggest that these may have something interesting to teach us about the process of death and our human psyche.

When viewed dispassionately they seem to push the limits of the mind-brain relationship as understood by our current state of comprehension and compel us to develop a newer, broader, more encompassing 'science' of consciousness.

# Closing Thoughts

Up until the close of the twentieth century, physics dealt with *linear* systems or formulaic approximations to generalized situations. In linear systems, of which there are many in Nature, small perturbations produce small effects. For example, the waveform of ripples produced by a pebble dropped onto a pristine lake surface can be described by linear mathematical equation, *ad infinitum*. However, the analysis of how they *break* on the beach requires mathematics on a very sophisticated and non-linear scale. This currently falls within the intriguing realm of Chaos Theory.

The evolution of the ripple pattern is so sensitive to the *initial* conditions or to subsequent disturbances that even a lonesome *butterfly*, flapping its wings deep in the woods of Lake Superior, might alter the *jet stream* blowing across the lake.

Consequently, the lake water could churn up into a *displacement trajectory* that ends up producing a massive wave at the shoreline instead of a weak ripple. This is also, why we are able to forecast the weather with reasonable accuracy only for the next few days, beyond which our predictions become increasingly less reliable. It is Chaos Theory that informs us why we can only hazard a guess via such non-linear models or barely compute the deterministic odds.

Careful spectroscopic analysis of cosmic microwave radiation has revealed that interstellar dust does contain a number of potentially biogenic molecules, which under Earth-like conditions could give rise to biologically significant compounds. In fact the Murchison meteorite, discovered in Australia back in 1969, was found to contain a number of amino acids remarkably similar in nature and relative quantity to those obtained by Miller, who as a grad student, decided to spark electricity through a sealed glass enclosure containing a gaseous mixture of methane, ammonia, hydrogen and water vapor.

This historic experiment alerted organic chemists to the origin of life as a probable chemical event, the inevitable product of deterministic forces and chemical reactions. It began through the spontaneous formation and interaction of small organic molecules widely distributed throughout the universe.

Given the unique physico-chemical conditions that prevailed on prebiotic Earth, these molecules were caught in a reaction spiral of increasing complexity eventually giving rise to the nucleic acids, proteins and other enzyme systems that dominate life processes today.

An important conclusion that emerged from these observations was the clear congruence, which seems to exist between proto-metabolism – the set of chemical reactions that first initiated life – and metabolism – the set of chemical reactions that support and maintain life today.

Organisms as different as bacteria, worms, corn plants and mammals contain similar proteins and nucleic acids, much closer than could possibly be accounted for by luck or the vagaries of chance. They enforce the inevitable conclusion that these molecules and by extrapolation, all organisms throughout the living world are not only intimately related to each other, but must be derived from a common ancestor.

The general consensus among modern-day Geochemists favor the origins of life in a hot, roiling prebiotic Earth, bombarded by intense bolts of lightning and saturated by the mutagenic ultraviolet radiation streaming down from the youthful Sun. Also favoring a hot cradle for life is the recent discovery of deep oceanic hydrothermal vents, which harbor numerous strains of Archaebacteria.

Such bacterial forms are the simplest progenitors of life on Earth. Their progeny have now colonized almost every possible kind of habitat, from the balmy shelter of the human gut to the salty brine of tidal pools. The richest source of bacteria is the soil, where these invisible organisms, usually no more than a few hundred-thousandths of an inch in size, silently accomplish the all-important decomposition of dead plants and animals, thus helping to recycle the vital constituents of life.

Fossil traces of these ubiquitous organisms, obtained from specially layered rock formations dubbed *stromatolites,* aggregate into superimposed mats, each consisting of a different kind of bacterial species. The top layers of the colony convert sunlight energy to produce nutrients, which cycle through the lower layers upon their demise. Some of these formations date back to circa 4.2 billion years ago (bya).

Even though *stromatolites* and microfossils do not appear any different today then a few billion years ago, this obvious 'stagnation' is quite misleading. Events of cardinal importance took place in the very crypts of these rocky formations, gearing up for the great explosion (Cambrian) of life forms that burst forth around 600 mya.

Detailed physical analysis of fossil carbon deposits (kerogen) obtained from these stromatolites show an excess of carbon isotope 12 versus isotope 13. Such 'enrichment' traditionally signals biological carbon assimilation i.e. it points to the presence of cellular metabolism deep within living organisms.

The scientists concluded that the common ancestor of life probably appeared on Earth sometime between 4.2 and 4.0 bya.

But... why? Why did the Earth 'choose' not to stay abiotic or prebiotic like the other 95 worlds we have discovered so far in our solar system?

What prompted life to take hold here, the third rock from the Sun?

The dynamic equilibrium that seems to exist within a living stromatolite also stabilizes other parts of the Biosphere. Even the simplest of fields or ponds are multi-factorial systems stitched into intricate networks by dynamic interactions among the various plants, animals, fungi and microorganisms they contain. Such systems, in turn, join to create larger and more complex grids, eventually forming a single, gigantic web, if you will, that envelopes the entire Earth.

However, the Biosphere is not just a pellicle of living matter. According to the *Gaia* Hypothesis, proposed in the 1970's by Dr. Lovelock, the Earth is a living organism that automatically regulates its environment to ensure its compatability to life.

The Biosphere is intimately linked to the physical Earth by myriad reciprocal pathways that draw from and act on the crust, oceans and atmosphere around it. Life and Earth are so undissociable that some scientists view them as joined into a sort of planetary superorganism, held together by a network of cybernetic relationships.

To take this analogy one step further, this biospheric entity is embedded in a cosmic *milieu* where, in keeping with Dr. Teilhard's vision, mind and matter coexist in elementary form from the beginning of the Universe and are jointly driven toward increasing complexity by the combined and complementary powers of these two forms of energy. Life emerged naturally from this 'complexification' and went on to weave an increasingly elaborate fabric of living organisms around our planet. The evolution of this 'biosphere' culminated in the appearance of humankind and consciousness.

Surprisingly, Teilhard's philosophy has gained progressive acceptance among a number of physicists and cosmologists. In 1974, Brandon Carter proposed the so-called "Anthropic Principle" which comes in two flavors. In its 'weak' form, it allows the existence of 'other' universes *without* intelligent life forms that cannot be 'perceived' as they do not contain human beings.

The 'strong' version asserts that the universe must be such as to produce intelligent life. The 'participatory' version of this Principle, put forward by John Wheeler, asserts that observers are actually *necessary* to bring the universe into being, a definition that seems to call for some sort of *retroactive creation* by the human mind *after* it emerged. In other words, according to Wheeler, we humans make-up stuff and create our own sense of Reality.

Such speculations have become a hot topic of contemporary discussions. A new and improved Anthropic *Cosmological* Principle, marshalling evidence from history, philosophy, religion, biology, physics, astrophysics, cosmology, quantum mechanics and biochemistry in support of their general points of view was proposed in 1986 by Barrow & Tippler et al. According to them, the evolving cosmos of modern science can be combined with an ultimate meaningfulness to Reality, within the basic framework of Teilhard's theory.

I conclude this book with a quote from Nobel laureate Steven Weinberg:

"It is almost irresistible for humans to believe that we have some special relation to the universe, that human life is not just a more-or-less farcical outcome of a chain of accidents reaching back to the first three minutes, that we were somehow built in from the beginning…"

Hmmm!

Meet up with you next in *Cognitive Optometry 3*, where I shall explore the Neuro-Chemistry of the Human Brain.

# END

# Further Reading:

In Search of Memory : Eric R. Kandel

Synaptic Self : Joseph LeDoux

The Mind & the Brain : Jeffrey M. Schwartz

How the Mind Works : Steven Pinker

The Perennial Philosophy : Aldous Huxley

The Dream Drugstore : J. Allan Hobson

The Varieties of Religious Experience : William James

Consciousness : Susan Blackmore

A Universe of Consciousness : Gerald Edelman

The Quest for Consciousness : Christof Koch

Cognitive Neuroscience : Gazzaniga, Ivry & Mangun

The Central Nervous System : Per Brodal

The Private Life of the Brain : Susan Greenfield

Hidden Dimensions : B. Alan Wallace

# About the Author

Dr. Nazir Brelvi is an Optometric Physician in a busy private practice. Apart from his primer on Cognitive Optometry back in 2007, of which this current book is a continuation, Nazir has authored several riveting science-based adventure books that explore pre-human evolution between three million and 1.4 million years ago.

Dr. Brelvi graduated from SUNY: State College of Optometry in New York City and has traveled extensively in India, Africa, Europe and South America.

Realizing in 2007, that some school children in the greater Hackettstown/ Mount Olive area were not receiving adequate vision care,

Dr. Brelvi founded the KidSEE 20/20 Foundation, an innovative non-profit organization that offers eye exams to kids at nominal fees.

Dr. Brelvi lives in Panther Valley of Allamuchy, NJ with his wife, Sarah and their little Shih-Tzu, Suki.